# Headline Series

No. 300     **FOREIGN POLICY ASSOCIATION**     Fall

# Environmental Scarcity and Global Security

## by Thomas F. Homer-Dixon

Cover Design: Ed Bohon     $5.95

*The American Academy of Arts and Sciences has
contributed to this issue of the HEADLINE SERIES in the
interest of fostering a better understanding of the environmental aspects of national and international security.*

# The Author

THOMAS HOMER-DIXON is coordinator of the
Peace and Conflict Studies Program at the
University of Toronto and assistant professor
in the Department of Political Science. He is
also codirector of an international research
project on Environmental Change and Acute
Conflict, jointly organized by his program and
the American Academy of Arts and Sciences
in Cambridge, Massachusetts. He received his
doctorate in political science from MIT and

has written on a broad range of topics—from arms control to environmental policy to the philosophy of social science. His most
recent article is "On the Threshold: Environmental Changes As
Causes of Acute Conflict," *International Security* (Fall 1991).

---

## The Foreign Policy Association

The Foreign Policy Association is a private, nonprofit, nonpartisan
educational organization. Its purpose is to stimulate wider interest
and more effective participation in, and greater understanding of,
world affairs among American citizens. Among its activities is the
continuous publication, dating from 1935, of the HEADLINE SERIES.
The author is responsible for factual accuracy and for the views ex-
pressed. FPA itself takes no position on issues of U.S. foreign policy.

---

HEADLINE SERIES (ISSN 0017-8780) is published four times a year, Winter, Spring,
Summer and Fall, by the Foreign Policy Association, Inc., 729 Seventh Ave., New
York, N.Y. 10019. Chairman, Michael H. Coles; President, R.T. Curran; Editor in
Chief, Nancy L. Hoepli; Senior Editors, Ann R. Monjo and K.M. Rohan. Subscription
rates, $15.00 for 4 issues; $25.00 for 8 issues; $30.00 for 12 issues. Single copy price:
$5.95. Discount 25% on 10 to 99 copies; 30% on 100 to 499; 35% on 500 to 999; 40%
on 1,000 or more. Payment must accompany all orders. Postage and handling: $2.50
for first copy; $.50 each additional copy. POSTMASTER: Send address changes to
HEADLINE SERIES, Foreign Policy Association, 729 Seventh Ave., New York, N.Y.
10019. Copyright 1993 by Foreign Policy Association, Inc. Design by K.M. Rohan.
Printed at Science Press, Ephrata, Pennsylvania. Fall 1992. Published June 1993.

Library of Congress Catalog Card No. 93-070280
ISBN 0-87124-152-8

# Introduction

By the year 2025, the planet's population will have soared from today's 5.5 billion to a figure probably over 8 billion. Global economic output—the total production of goods and services traded in markets around the planet—will also have skyrocketed from around $25 trillion in today's dollars to nearly $60 trillion. In comparison, the United States, a populous and prosperous country by any standard, has in 1993 just over 250 million people, who produce about $5.5 trillion in goods and services.

Largely as a result of these two global trends of rising population and output, environmental change will take place much faster, and be much more widespread and

Portions of this HEADLINE SERIES have appeared in the author's article, "On the Threshold: Environmental Changes As Causes of Acute Conflict," *International Security* (Fall 1991), pp. 76–116; in his chapter, "Physical Dimensions of Global Change," in *Global Change: Environmental Challenges and International Responses*, Nazli Choucri, ed. Cambridge, Mass., MIT Press, 1993; and in the article he coauthored with Jeffrey Boutwell and George Rathjens, "Environmental Change and Violent Conflict," *Scientific American*, February 1993.

severe. Around the planet, human beings will face a steady decline in the total area of high-quality agricultural land. Many of the planet's remaining virgin forests will vanish along with the wealth of species they shelter. Numerous rivers, aquifers* and other water resources will become exhausted and polluted; key fisheries will collapse; further ozone depletion* will occur in the stratosphere; and global warming* may cause significant climate alteration.

These environmental changes can be thought of as "scarcities" of vital resources such as soil, water and a stable climate. Within the lifetimes of our children and grandchildren, these environmental scarcities may cause widespread social disorder and violence, including war, revolution, ethnic violence, riots and coups that topple established governments. Many people, experts and laymen alike, find it difficult to imagine exactly how this might happen. Here are two examples from today's world.

In the Indian subcontinent, environmental scarcity has contributed to vast migration and, in turn, horrible ethnic violence within countries. "For us, it is a question of survival," said Mugha ul-Khand, a farmer in the Mymenshingh district of Bangladesh, in early 1991. "We will go anywhere because every year our landholding is shrinking; our families are growing. My father had 24 bighas [approximately 8.4 acres] of land; now my sons have 2 bighas each. What can you grow on 2 bighas? In the future, we may have nothing. Yes, we will be prepared to go anywhere, to Assam, if necessary, if we can get land and live with dignity."

Since the 1950s, millions of people have fled appalling population pressures, land scarcity and flooding in the region that is now Bangladesh. They have poured into neighboring areas of India, especially the state of Assam, whose population of 22 million is about 7 million larger than it would have been without these migrants and their descendants. (In proportion to the original population, it is as if California had to absorb an extra 14 million people, increasing the state's population to over 44 million from today's 30 million.) The migrating

---

*All starred words appear in glossary on page 32.

Bengalis are not eagerly welcomed. Members of the Lalung tribe, for instance, have long resented Muslim Bengali migrants, and they accuse the newcomers of stealing the area's richest farmland. In early 1983, violence finally exploded during a bitter election for federal offices in the state. In the village of Nellie, Lalung tribespeople used machetes, knives and spears to hack to death nearly 1,700 Bengalis in one five-hour rampage.

## *Frictions Between Nations*

Scarcities of environmental resources can also heighten tensions between countries. On January 13, 1990, for example, Turkey began filling the giant reservoir behind the new Ataturk Dam in eastern Turkey. For one month Turkey held back the main flow of the Euphrates River, which cut the downstream flow in Syria to about a quarter of its normal rate. By early in the next century, Turkey plans to build a huge complex of 20 dams and irrigation systems along the upper reaches of the Euphrates. This will reduce the annual average flow of the Euphrates within Syria by more than a third. Syria is already desperately short of water: almost all water for its towns, industries and farms comes from the Euphrates, and the country has been chronically vulnerable to drought. Syria's population growth rate, at 3.6 percent per year, is one of the highest in the world, which gives a further push to the country's demand for water.

Turkey and Syria have exchanged angry threats over this situation. Syria gives sanctuary to guerrillas of the Kurdish Workers' party (PKK), which has long been waging an insurgency against the Turkish government in the eastern part of the country. Turkey suspects that Syria might be using these separatists to gain leverage in bargaining over Euphrates water. Thus, in October 1989, then Prime Minister Turgut Ozal said that Turkey might cut off the river's water if Syria did not restrain the PKK. Although he later retracted the threat, the tensions have not been resolved, and there are currently no high-level talks on water-sharing.

Are there other indicators that environmental scarcities can precipitate conflict? Until recently, a good answer to this question has been impossible because of lack of evidence. But in the last three years, various researchers have gathered enough information to reach a disturbing conclusion: environmental scarcities are already contributing to violent conflicts in many parts of the developing world. Moreover, these conflicts may be the early signs of an upsurge in violence in the coming decades—especially in poor countries—that is caused or aggravated by environmental change.

This HEADLINE SERIES therefore addresses these questions: What are the principal large-scale environmental changes that may confront the human species over the next decades? How might these changes affect national and international security? How are these changes related to the level of technological development, population growth and institutions and culture? And can societies limit or adapt to these changes in order to lessen threats to human security?

First, though, what is meant here by "security"? Within the field of international relations, the concept has traditionally meant safety from violence and military threats to one's society from outside one's boundaries. But with the disintegration of the Soviet Union, which was the immediate enemy facing the United States and other Western societies, and with increasing evidence that all peoples face daunting social and environmental problems, many have suggested that the concept of security should be broadened. It might, for example, be expanded to include nonmilitary threats to national and personal well-being, covering threats to health, prosperity and general quality of life.

But such a broadened concept of security brings together too many different and complex issues. This book, therefore, uses the more traditional definition: it deals with the relationship between environmental change and potentially violent conflict, both within and between countries.

What are the possible links between environmental change and threats to security? Some experts have proposed that environmental change may shift the balance of power between states either regionally or globally, causing instabilities that could lead to war. For example, some claim that global warming could devastate American crop production and hurt the economy, which could eventually contribute to a weakening of U.S. military power and make the United States less able to defend its friends and allies from attack. Another possibility is that global environmental damage of all kinds—from soil degradation to ozone depletion—might increase the gap between rich and poor nations, with the poor then violently confronting the rich for a fairer share of the world's wealth. Imagine a world in which, for example, environmental hardship helps a radical group seize the government of a powerful poor country; the new leaders might then direct popular anger and military attacks against rich countries in order to increase their support at home.

### 'Free-riders'

Severe conflict may also arise from frustration with countries that do not go along with agreements to protect the global environment or that are "free-riders," letting other countries absorb the costs of environmental protection. Over the next decades, for instance, China will be rapidly raising its emissions of gases that cause global warming. Might the rest of the world threaten economic sanctions, or even go to war, to prevent one country from wrecking the global environment?

Warmer temperatures could lead to contention over new ice-free sea-lanes in the Arctic and over more-easily harvested resources in the Antarctic. Bulging populations and land stress, as seen in Bangladesh, may produce waves of environmental refugees spilling across borders and disrupting relations among ethnic groups. Countries such as Turkey and Syria may fight among themselves because of dwindling supplies of water and the effects of upstream pollution. A sharp decline in food-crop production and grazing land could lead to conflict between nomadic tribes and sedentary

7

farmers, as is already happening in many parts of the Sahel in Africa. Moreover, if environmental degradation* makes food supplies increasingly tight, exporters could use food as a weapon. For instance, in southern Africa, countries that normally have food surpluses, such as South Africa or Zimbabwe, might be tempted to force hungry neighbors to alter their foreign and domestic policies.

Environmental change could in time cause a slow deepening of poverty in poor countries, which might open bitter divisions between classes and ethnic groups, corrode democratic institutions, and spawn revolutions and insurgencies. In general, many experts have the sense that environmental degradation will ratchet up the level of stress within nations and within international society, increasing the likelihood of many different kinds of conflict—from war and rebellion to trade disputes—and undermining possibilities for cooperation.

Which of these scenarios are most plausible, and why? The following pages explore why environmental issues have become prominent in recent years and provide a framework for understanding them. The piece continues with an overview of key environmental variables, including population growth, energy consumption, climate change, ozone depletion, deforestation, loss of agricultural land, decline in water supplies, depletion of fish stocks and loss of biodiversity.* Then the likely links between environmental change and conflict are identified and illustrated with several case studies. These are followed by a discussion of how much societies can adapt as environmental problems become severe, and the book concludes with recommendations for action by both rich and poor countries.

The case studies reflect the results of a three-year research project on "Environmental Change and Acute Conflict," organized jointly by the Peace and Conflict Studies Program at the University of Toronto and the American Academy of Arts and Sciences in Cambridge, Massachusetts. This project brought together a team of 30 researchers from four continents.

The research found that poor countries are likely to be affected sooner and more harshly by environmental scarcity than rich countries. In general, they do not have the financial, material or intellectual resources of highly industrialized countries. For instance, they find it harder to build the new dams, irrigation canals and roads to adapt to shifts in rainfall from one region to another that might be caused by greenhouse warming. Moreover, the economic and political institutions of poor countries are often weaker and less flexible. They often have little expertise to manage resources; their governments may not have the money to "buy off" or placate competing groups whose interests are disrupted by scarcities; and common standards of responsibility to the community are often only weakly held by leaders and citizens. Poor societies therefore have less ability to buffer themselves from environmental scarcity and the social crises it can cause.

### Unforeseen Thresholds

The project's research also suggests that future violence arising from environmental change will generally not follow the age-old pattern of conflicts over scarce resources, where one group or nation tries to seize the water, oil or minerals of another. This is partly because some environmental resources—such as the climate and the ozone layer—are held in common; in other words, they are not privately owned by either individuals, groups or corporations. It is also because of the often insidious social effects of environmental scarcity, such as slow population displacement and economic disruption, that can in turn lead to clashes between ethnic groups and rebellion. While these types of conflict may not be as conspicuous or dramatic as wars between countries over scarce resources, they may have critical implications for the security interests of rich and poor nations alike.

While the last decades have seen increasing environmental damage around the globe, for the most part this change has progressed slowly, one small change at a time. Why, then, are people suddenly paying attention to environmental issues?

Clearly, the end of the cold war has given the public and its leaders a chance to think about other things. But there is another factor at work: during the last 10 years there has been a genuine shift in experts' perceptions of global environmental problems. They used to perceive environmental systems as relatively stable and resilient in spite of the fact that humans were harvesting resources and dumping wastes on a massive scale. They thought that these systems—such as the earth's climate and the interlinked networks of animals, plants and energy sources that make up the earth's fisheries, forests and agricultural lands—would change only slowly in response to human insults. But now scientists widely believe the behavior of these systems is often quite unpredictable and unstable. In trying to describe these systems mathematically, using equations relating the values of certain variables (e.g., energy consumption) to other variables (e.g., population growth), they discovered that many of the mathematical relationships between the variables are sharply "nonlinear." This means that the slope of the function relating the value of variable X to variable Y changes abruptly as the value of X passes a certain point, which is often called a "threshold" (see Figure 1). In practical terms, this means that it may be much easier than was thought to push an environmental system from one equilibrium state to a very different equilibrium state.

In 1987, geochemist Wallace Broecker commented on recent polar ice-core and ocean-sediment data:

> What these records indicate is that earth's climate does not respond to forcing [i.e., human interventions in the environment] in a smooth and gradual way. Rather, it responds in sharp jumps which involve large-scale reorganization of earth's system…. We must consider the possibility that the main responses of the system to our provocation of the atmosphere will come in jumps whose timing and magnitude are unpredictable.

## Figure 1. Nonlinear Functions and Threshold Effects

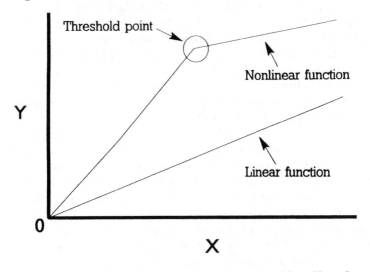

Constant pressure may not have a noticeable effect for a long period. But sooner or later the system's resilience or buffering capacity is gone and even a small additional pressure nudges it across a critical threshold.

A dramatic example of such a threshold effect in a complex environmental system is the appearance of the Antarctic ozone hole in the mid-1980s. The scientific models of ozone depletion used to that point had, for the most part, assumed a roughly linear relationship between emissions of chlorofluorocarbons (CFCs)* and ozone depletion; in other words, it was assumed that ozone depletion would increase in steady, constant steps as CFC emissions increased. (CFCs are now used mainly in air conditioners and refrigerators, and they are released into the atmosphere when these machines are repaired or scrapped.) Atmospheric scientists had not even remotely anticipated the ozone-destroying process that occurs on the surface of ice crystals in the stratosphere when there is a critical combination of temperature, light, water, nitrogen and CFCs. If the conditions are right, it turns out, this destruction can occur at lightning speed, stripping the ozone from mile-thick layers of the stratosphere in a matter of days. The Antarctic ozone hole is startling evidence of the instability of environmental systems in response to human inputs, of the ability of humans to affect significantly the ecosystem* on a global scale, and of humans' inability to predict exactly how the system will change.

This new perception of the nature of the environmental system has percolated out of the scientific community to influence the broader public's view of environmental problems. This in turn has affected the views of the policymaking community in governments. Scientists, political leaders and the general public are beginning to interpret data about environmental change in a new light: constant, step-by-step degradation of environmental systems is not as tolerable as it once was, because of uncertainty about where and when a threshold might be crossed, resulting in a move to a radically different and perhaps very undesirable system.

This new perception is rooted in a more mature understanding of natural systems and the global damage humans are inflicting on these systems. This understanding will likely endure and, along with it, strong concern about the environment. Over the next 30 years there will be no shortage of ever more ominous environmental signals from our planet. Even if there are no dramatic, nonlinear shifts in the ecosystem (though their probability may be quite high), environmental problems will remain prominent on scientific, policy and public agendas.

# Where Will the World Be in 2025?

The environmental problems facing humankind might seem overwhelming. They are large-scale, long-term and poorly understood. They strike directly at our most intimate links to the biosphere,* which is the thin shell of life—only about five miles thick—covering the planet like the skin of an apple. These links include the ability to get the food and water needed for survival and the stability of our societies.

### Key Variables and Relationships

Although today's environmental problems seem frightening and often unsolvable, it is important to avoid slipping into simpleminded "environmental determinism"; in other words, one must not assume that the environment surrounding us determines, inescapably and inevitably, people's behavior and level of happiness. One must remember that societies are often very flexible, that human beings are often very creative, and that many factors often permit great variability and adaptability in human environmental systems.

# Figure 2. Main Variables and Causal Relations

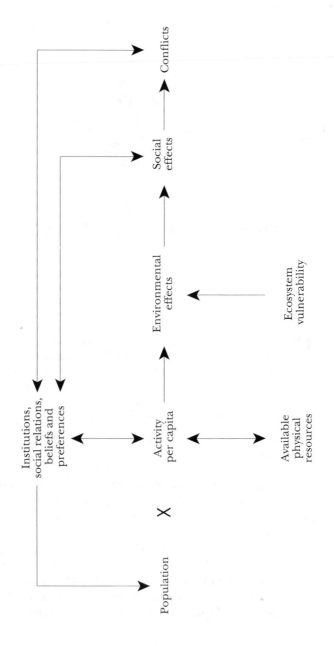

Some of these factors are identified in Figure 2. It shows that the environmental effects from human activity in a particular ecological region are a function of two main variables: first, the product of total population in the region and physical activity per person (which depends on the range of technologies people use in the region); and, second, the vulnerability of the ecosystem in that region to those particular physical activities. Say one is interested, for example, in calculating the total greenhouse effect of car transportation in the United States. Since car engines produce the greenhouse gas* carbon dioxide, multiply the U.S. per capita emissions of carbon dioxide from driving by the number of people in the United States to get the total emission of carbon dioxide from driving. Next, get an estimate of the vulnerability of the ecosystem, in this case the earth's climate, to this particular quantity of pollution. With this estimate in hand, the total effect on the earth's climate of all automobile transportation in the United States can be estimated.

In Figure 2, activity per capita is, in turn, a function of available physical resources in the region (which include nonrenewable resources such as minerals, and renewable resources such as water, forests and agricultural land) and social and psychological factors, including institutions, social relations, preferences and beliefs. The figure also shows that environmental effects, such as the degradation of agricultural land, might cause certain types of social effects, such as the large-scale migration of people out of the affected region. This could in turn lead to various kinds of conflict, such as the ethnic conflicts in Assam that resulted when newcomers came in contact and began competing with local groups. There are important feedback loops from these social effects and conflicts to the social and psychological factors at the top of the figure, and then back to activity per capita and population. For example, migration could alter relationships among classes and ethnic groups in a society, which might in turn affect its economic activity.

The social and psychological factors at the top of the diagram are particularly important. They are broad and complex. They include the degree of inequality of land and wealth distribution in the society; the system of property rights and markets that either encourages people to produce goods and services within the economy or discourages them from such activity; family and community structures; historically rooted patterns of trade with other societies; the distribution of power among countries in the international system; and beliefs about the relationship between people and nature.

Without a full understanding of these factors one cannot begin to grasp the true nature of the relationships between human activity, environmental change and its social effects. These factors help determine the form and rate of the physical activities in a society, and the vulnerability, adaptability and flexibility of the society when faced with environmental stress.

For example, at first glance a sudden drought and collapse of grain output in a rural area of Africa might seem to mean inevitable hardship and starvation for farming families. But in many areas where drought is common, families and villages have developed traditional ways of lessening the impact. Grain might be stockpiled in anticipation of a dry year; communities across a region may have long-standing agreements to help villages that cannot grow enough food by sharing surpluses from those that can; or the traditional agricultural system may be designed to shift rapidly to crops that are more tolerant of drought. These are sets of beliefs and cooperative arrangements shared by individuals and communities. A good understanding of such factors will prevent falling into the trap of environmental determinism.

### Trends and Projections

This section provides information on current trends and projections for nine key environmental variables. In most cases, the rate of change of the variable's value has increased sharply this century. For example, while in 1900 about 10 million people were added to the world's total population, nearly 90 million were added in 1992. This population growth has combined with dramatically higher per capita consumption of material goods—especially in North America, Europe and Japan—to produce staggering jumps in energy consumption, carbon emissions, water consumption, fish consumption, land degradation and deforestation.

Because many of the statistics cited here were produced by combining or averaging figures for numerous regions and societies, they hide important regional differences. Moreover, societies differ in the value they assign to a given environmental good or service, so some societies will be more severely affected than others by an equivalent physical change in the environment. For instance, in most of Africa about 60 percent of the work force has jobs in agriculture, whereas in the most highly industrialized countries the figure is usually below 5 percent; a certain loss of agricultural soil will therefore hurt fewer people in industrialized countries and will be less of a concern to the society as a whole.

Despite these qualifications, the figures and trends offered below give a quick overview of the extent and nature of human activity and its environmental effects today and in the year 2025.

- **Population Growth**

Figure 2 shows that population size can be a key variable driving environmental change. Sometimes it does not damage the environment, but often this growth—when combined with certain social structures, technologies and patterns of consumption—makes environmental degradation worse, as will be evident in the case studies.

In the 1960s, certain experts rang the alarm on global population growth, which at the time was well over 2 percent a year. (While this may not seem like a frightening figure, at that rate the earth's population doubles in size every 35 years.) These experts claimed that climbing human numbers would mean critical shortfalls of food, space and other resources; huge, violent cities without work for masses of poor; and greater misery for future generations. But during the 1970s and early 1980s, with the aid of economic growth, education and active family-planning programs, average family size dropped dramatically in many poor countries, from six or seven children to three or four. Since then, though, experts have discovered that it is much more difficult to convince parents to forgo a further one or two children to bring family size down to a replacement rate of about two children per couple. As a result, the population growth rates of some of the world's most crowded countries—including India and Bangladesh, which together account for almost 20 percent of the world's population—are hardly declining. India's rate has leveled off at around 2.0 percent (about 18 million people) per year, while Bangladesh's has stalled at around 2.6 percent (about 3 million) per year.

These developments have recently led the United Nations to revise upward its mid-range estimate of the globe's population in 2025 from 8.2 to 8.5 billion. The UN has also raised its estimate of world population when it stabilizes (predicted to occur toward the end of the twenty-first century) from 10.2 to 11.3 billion, over twice the size of the planet's current population.

Demographers, who are experts in population statistics, have long assumed that poor countries would pass through a "demographic transition." This transition occurred in the currently rich countries in the nineteenth and twentieth centuries, when a decline in deathrates was eventually followed by a compensating decline in birthrates. The change in deathrates was mostly produced by improved nutrition and basic sanitation. The change in birthrates is thought to have resulted from increased wealth and certain social changes, such as higher literacy rates and social status for women. But if certain developing countries cannot maintain a steady growth in social and economic wealth, they may not undergo a demographic transition.

- **Energy Consumption**

Total commercial energy consumption in 1988 was roughly equivalent to 8 billion metric tons of oil, an amount that would fill a huge 1.3-mile cube. The 1988 figure for world energy consumption was an increase of nearly 4 percent over 1987; and during the past two decades, energy consumption's steady climb was broken only in the early 1980s. The U.S. Department of Energy expects the world rate of energy consumption to increase at up to 2 percent annually through the year 2000, which gives a doubling time of about 35 years. Energy consumption per person in many rich countries is 30 or more times greater than that in poor countries. But the energy consumption per person in rich countries has not increased much in the last decade, while in some rapidly developing countries this figure is rising very fast. As industrialization, electrification and transportation networks expand in poor countries, their appetite for energy may grow up to twice as fast as that in the rich countries.

Currently, oil makes up 38 percent of global commercial energy consumption; coal, 30 percent; natural gas, 20 percent; hydropower, about 7 percent; and nuclear power, 5 percent. As accessible and cheap petroleum reserves disappear, these percentages will change. In the first decades of the next century, many experts predict that oil consumption will drop, while natural gas, nuclear power, nonconventional sources (such as solar, wind and tidal power), and perhaps coal will have to fill the gap. In addition, many people in poor countries depend for cooking and heating on noncommercial sources of energy, such as wood gathered in forests and straw from fields. In large areas of the world, especially Africa, the Indian subcontinent and China, wood is increasingly hard to find and gatherers have to walk miles from their villages to find the day's fuel.

The level of global energy consumption in 2025 will depend on a huge array of factors, including whether energy prices are adjusted so they reflect the costs of pollution and global warming to our societies, whether new technologies allow more-efficient production and use of energy, and whether diplomats can reach international agreements to limit the release of carbon dioxide. Because of this uncertainty, experts' estimates of global energy demand in 2025 run from a low of the equivalent of 9 billion metric tons of oil to a high of 26 billion tons.

- **Climate Change**

As is now widely understood, human activities—especially in industrialized countries—release a number of gases (mainly carbon dioxide, CFCs, methane and nitrous oxide) that slow the escape of infrared radiation (heat) from the surface of the earth into space. In rough terms, one can say that (given a certain amount of incoming sunlight) the more of these gases present in the atmosphere, the higher the average temperature at the surface of the planet. In fact, if it were not for the naturally occurring greenhouse effect, the average temperature on earth would be about 60 degrees Fahrenheit lower than it is now and the planet would not be able to support most forms of life.

But the scientific story is much more complicated than this, because there are many feedback loops within the climate system that may worsen or diminish the effects of human activity. For instance, scientists believe that global warming will increase the amount of water vapor in the atmosphere. This will increase the number and extent of clouds, since clouds are condensed water vapor. But scientists are uncertain whether more clouds will trap further heat like a blanket (a positive feedback) or reflect sunlight like a mirror (a negative feedback).

Over the last few years, though, a number of experts have reached a rough consensus: assuming no major changes in the trend of human emission of greenhouse gases, the earth will warm an average of nearly 2 degrees Fahrenheit by 2025 and 5 degrees by 2100. This may not seem like much, but the earth has warmed only some 9 degrees since the coldest period of the last ice age, about 18,000 years ago. Moreover, the predicted rate of increase during the next 100 years will be over 0.5 degrees per decade, which is far faster than any climate change in recorded history.

What could be the impact of such a warming? At the moment, the complicated computer models of the atmosphere that scientists use to estimate future global warming cannot accurately predict changes in patterns of rainfall, storms and soil moisture for specific regions. For instance, surface features like mountain ranges have a great effect on winds and rainfall, but are not yet modeled properly. However, some things are known: temperature increases toward the poles of the planet will be much greater than at the equator; sea levels will rise about two inches per decade (mainly because seawater expands when it is warmed); coastal areas will generally get more rain; and interiors of continents will become drier. According to some estimates, by 2030 central North America could warm from 3 to 7 degrees in winter and 3 to 5 degrees in summer and experience a 15 to 20 percent drop in soil moisture. Since soil moisture is a key variable for farmers, this change could seriously affect grain output in the United States and Canada.

Finally, when considering the social impact of climate change, the most important result of a rise in average temperature may be its effect on the likelihood of "extreme" environmental events. While a 2 to 5 degree average global warming might not seem too significant for food production, it could cause a sudden increase in the chance of crop-devastating droughts, floods, heat waves and storms, even if it does not force the climate system to a completely new equilibrium.

- **Ozone Depletion**

The Antarctic ozone hole contributes to the general depletion of ozone over a wide area of the Southern Hemisphere. Each southern spring, the hole forms inside a circular pattern of wind called the circumpolar vortex; as summer approaches, this vortex breaks up, and ozone-depleted air moves northward from Antarctica. Although Antarctica's pattern of chemical and atmospheric events is not exactly duplicated over the North Pole, during the last three years scientists have found ominous signs that rapid ozone depletion may eventually appear there too. In addition, while the situation over the poles is perhaps the most urgent, ozone depletion is occurring around the planet as CFCs move into the upper atmosphere. In early 1991, the U.S. Environmental Protection Agency (EPA) announced that satellite data showed an ozone decrease of 4.5 to 5 percent in the last decade over the United States, and 8 percent over Northern Europe; losses in the Southern Hemisphere (outside Antarctica) were on average 2 percent higher than those in the Northern Hemisphere.

Industrialized countries have already released immense quantities of CFCs into the atmosphere. It takes on average about 10 years for CFC molecules to migrate from the ground to the middle stratosphere, about 10 to 15 miles above the earth's surface. Once there, sunlight breaks down these molecules, and the liberated chlorine can catalyze the destruction of ozone molecules for decades before falling back into the lower atmosphere. So, with the CFCs released to date, a dramatic thinning of the ozone layer over the coming decades is probably inevitable. It is very hard to give precise estimates of future loss; current mathematical models, based on today's knowledge of the atmosphere, have generally underestimated the rate of ozone loss. However, the EPA now predicts worldwide depletion of 10 to 12 percent by 2010, and there may be a 15 to 20 percent depletion by 2025.

Recent research suggests that a 1 percent decrease in stratospheric ozone produces about a 2 percent increase in the incidence of cancer-causing ultraviolet radiation on the surface of the earth, which in turn produces about a 3 percent increase in nonmelanoma skin-cancer rates. But the increasing rate of skin cancer may not be the worst problem: the harmful effects of increased ultraviolet radiation on crops, forests, ocean phytoplankton (which are at the bottom of the ocean food chain) and the health of people and livestock may be very grave. But research on these effects is still preliminary.

While greenhouse warming and ozone depletion have caught the public's attention over the last few years, environmental problems such as deforestation, soil degradation, depletion and degradation of water resources, and depletion of fish stocks deserve equal consideration. Such problems may, in fact, multiply the harmful effects of atmospheric change; and they merit immediate concern because they are already seriously threatening the well-being and stability of many developing societies.

● **Forest Degradation**

Estimates of damage to tropical forests vary widely, since there are different kinds and degrees of forest degradation. Furthermore, logged forests sometimes can recover through planting and natural regeneration, which also tends to blur category boundaries. Finally, satellite images are far less useful than commonly thought in allowing researchers to assess the extent of forest damage. The images usually have to be supported by detailed inspections on the ground.

Despite these problems, recent estimates by the World Resources Institute (WRI) suggest there has been a rapid increase in the rate of tropical forest depletion since the 1970s. Whereas the UN Food and Agriculture Agency (FAO) estimated in 1980 that the world was losing 11.4 million hectares of tropical forest annually, the new WRI study (based on more recent FAO data) says the figure may be as high as 20.4 million hectares, an area about the size of New York, Connecticut, Massachusetts, Vermont and New Hampshire combined. (There are about 260 hectares in a square mile.) The total area of closed tropical forest on earth, which means forest where the branches of the trees largely obscure the ground when viewed from the air, is estimated to be around 1.2 billion hectares, or about the area of the United States and Mexico together.

Particularly affected by these increased rates, according to the WRI, are the forests of Brazil, Costa Rica, India, Myanmar (formerly Burma), the Philippines and Vietnam. However, the Brazilians have responded that in their case the WRI figures are inflated, because the rate of forest degradation in Brazil has dropped dramatically since 1988 due to changes in government policy. Given that forest-degradation data are unreliable and quite susceptible to policy changes, it is hard to predict the state of the world's forests in 2025. But it seems safe to say that most of the remaining virgin tropical forests in Southeast Asia, South Asia and Central America will be gone, and the remainder will be concentrated in Zaïre and Brazil.

A close look at the Philippines reveals the speed and extent of forest loss. As recently as World War II, about half the area of the Philippine archipelago was forested. Since then, logging and the spread of farms has cut the virgin and new forest from about 16 million hectares to between 6 million and 7 million hectares. At the turn of the century, the Philippines had about 10 million hectares of virgin forest; now less than a million hectares remains, and it seems certain that almost all of this will be gone by early in the next century. The logging industry boomed in the 1960s and 1970s and, following the declaration of martial law in 1972, President Ferdinand E. Marcos handed out concessions to huge tracts of land to his cronies and senior military officials. Pressured to make payments on the country's foreign debt, the government encouraged log exports to the voracious Japanese market. Despite the toppling of Marcos and the more-aggressive concern for the environment of the government of President Corazon C. Aquino, the recent WRI estimates show the rate of damage to Philippine forests remains very high. Efforts to replant wide areas have generally failed, because of corruption and inefficiency in the agencies charged with the planting, because the hilly areas where the forest degradation is most critical have very thin soils, and because of the migration of landless peasants into the hills.

- **Agricultural Land**

Currently, total global cropland amounts to about 1.5 billion hectares, which is roughly twice the area of the contiguous 48 states. Optimistic estimates of potential cropland on the planet range from 3.2 billion to 3.4 billion hectares, but nearly all the best land has already been used. What is left is either less fertile, not sufficiently rain fed or easily irrigated, infested with pests, or harder to clear, plow and plant. Experts generally describe a country as land-scarce when 70 percent or more of potential cropland is under production. In Asia, which includes four of the world's five most populous countries, some 82 percent of all potential cropland is cultivated. While the percentages are lower in Africa and Latin America, the poor quality of the remaining land and its unfair distribution between rich and poor in these regions mean that the previously high rates of cropland expansion cannot continue.

For developing countries in general, cropland grew at just 0.26 percent a year during the 1980s, less than half the rate of the 1970s. More importantly, cropland per capita dropped by 1.9 percent a year. Without a major increase in the amount of cropland in developing countries, experts expect that the world average of 0.28 hectares per capita will decline to 0.17 hectares by the year 2025, given the current rate of world population growth. Large areas of land are being lost each year to a combination of problems, including encroachment by cities, erosion, depletion of nutrients, acidification, compacting and salinization and waterlogging from overirrigation. The geographer Vaclav Smil, who is generally very conservative in his judgments about environmental damage, estimates that 2 million to 3 million hectares of cropland are lost annually to erosion, with perhaps twice as much land going to expansion of cities and at least 1 million hectares abandoned because of salinity. In addition, about one fifth of the world's cropland is affected by desertification* (which includes wind erosion and changes in soil moisture). All told, he concludes, the planet will lose about 100 million hectares of arable land between 1985 and 2000, which is about the area of California and Texas together.

Smil gives a particularly startling account of the situation in China. From 1957 to 1977 the country lost 33 million hectares of farmland (30 percent of its 1957 total), while it added 21 million hectares of largely marginal land. He notes that "the net loss of 12 million hectares during a single generation when the country's population grew by about 300 million people means that per capita availability of arable land dropped by 40 percent and that China's farmland is now no more abundant than Bangladesh's—a mere one tenth of a hectare per capita!"

## • Water Supplies

The scarcity and pollution of freshwater supplies will be one of the chief resource issues of the next century. At the moment, humans withdraw about 3,500 cubic kilometers of fresh water a year from various sources (mainly rivers), chiefly for agriculture, and about 1,400 cubic kilometers are returned to these sources, often in a polluted condition. (For comparison, the total flow over Niagara Falls every year is about 90 cubic kilometers.) This consumption is growing at a rate of 2 to 3 percent a year. Total river resources at any one time amount to about 2,000 cubic kilometers, but because of the constant cycling of water between the atmosphere and surface of the earth, the amount available from rivers annually is probably closer to 40,000 cubic kilometers. But while these aggregate figures seem to say water is abundant, there are great differences between regions, and many areas—including much of Europe, large parts of the United States, the Ganges basin in India and the northwestern provinces of China—are using virtually all of their river runoff. In many dry poor countries, quick population growth may soon reduce per capita water availability to levels below those required to meet basic household, industrial and agricultural needs. If greenhouse-induced climate change causes large shifts in rainfall patterns, some of these regions may no longer face water shortages, while others may suffer ruinous drought.

It is possible to pinpoint certain regions where water crises are a virtual certainty by the year 2025. Table 1 on page 28 shows that the Middle East and certain parts of Africa are of particular concern because water is already very scarce, these regions' populations are growing rapidly, and water has long been a source of argument between certain groups and societies. For instance, the Nile River runs through nine countries, and downstream nations—Egypt and the Sudan, for example—may one day be especially vulnerable to upstream pollution or diversion of water, because they have dry climates and depend on irrigated agriculture. Other African rivers shared by several countries deserve attention, including the Zambezi and the Niger, which flow through eight and ten countries respectively, and the Senegal, which has been at the center of a recent crisis between Mauritania and Senegal. Depletion of aquifers may also be a source of conflict: Egypt and Libya, for example, see the Nubian aquifer that they share as a vital future source of water for huge agricultural zones. In the Middle East, some experts believe that the desire to secure the waters of the Jordan, Litani, Orontes and Yarmuk rivers contributed to tensions preceding the 1967 Arab-Israeli war. In addition, as described earlier, there is strong disagreement between Syria and Turkey over Euphrates water. Access to extremely limited underground-water resources is also an extra stress in the Israeli conflict with the Palestinians over the future of the West Bank of the Jordan River (see map on page 57).

- **Fish Stocks**

When considering the state of the world's fisheries, the critical issue is whether or not the current level of fishing can be continued indefinitely. In other words, can catches be "sustained" year after year? The FAO estimates the annual sustainable yield of the world's ocean and freshwater fisheries to be 100 million metric tons. Between 1950 and 1988, the quantity of fish brought ashore increased fivefold, from about 20 million to 98 million tons. (The U.S. catch every

## Table 1: Per Capita Water Availability in 1990 and in 2025

| | 1990 Per Capita Water Availability (m³/person/year) | 2025 Projected Per Capita Water Availability (m³/person/year) |
|---|---|---|
| **AFRICA** | | |
| Algeria | 750 | 380 |
| Egypt | 1,070 | 620 |
| Ethiopia | 2,360 | 980 |
| Kenya | 590 | 190 |
| Libya | 160 | 60 |
| Morocco | 1,200 | 680 |
| Nigeria | 2,660 | 1,000 |
| Somalia | 1,510 | 610 |
| South Africa | 1,420 | 790 |
| Tanzania | 2,780 | 900 |
| Tunisia | 530 | 330 |
| **THE AMERICAS** | | |
| Canada | 109,389 | 90,880 |
| United States | 9,940 | 8,260 |
| Haiti | 1,690 | 960 |
| Peru | 1,790 | 980 |
| **ASIA/MIDDLE EAST** | | |
| Cyprus | 1,290 | 1,000 |
| Iran | 2,080 | 960 |
| Israel | 470 | 310 |
| Jordan | 260 | 80 |
| Lebanon | 1,600 | 960 |
| Saudi Arabia | 160 | 50 |

The table lists selected countries where population growth will drive annual per capita water availability to 1,000 cubic meters or less per person by 2025. Many experts consider this the minimum water need for an efficient, industrialized nation. Several of the countries in the table will have less than 250 cubic meters per person available in 2025. Currently, no industrialized country uses so little water. Figures for the United States and Canada are provided for comparison. Source: Peter Gleick, Pacific Institute, Oakland, California.

year is about 6 million tons.) As the world approaches the threshold its fisheries can sustain, there is evidence—from Maine to Newfoundland to southern India—that fish stocks are collapsing. In 1987, the FAO commented: "The time of spectacular and sustained increases in fisheries catches is over.... Almost all important stocks...are either fully exploited or overfished. Many of the stocks of more highly valued species are depleted. Reef stocks and those of estuarine/littoral zones are under special threat from illegal fishing and environmental pollution."

By the year 2000, world demand for fish is predicted to rise to around 120 million tons, and annual requirements could exceed 160 million tons by 2025. Some of this demand will be met by expanding aquaculture (which involves raising fish in captivity), using species not normally eaten, and using fish that are currently thrown out. But as the thresholds of sustainability are passed, peasant and small-scale fishermen, especially in poor countries, will be hurt badly, as will poor people who depend on fish as a key source of animal protein.

- **Biodiversity**

"Biodiversity" refers to the range of plant and animal species on the planet. The estimates for the number of species on earth vary from 5 million to more than 30 million. Scientists are particularly uncertain about the range of species in the tropical forests, but they know that these areas contain a vast reservoir of genetic information. This is a priceless resource for the development of new crops, medicines and a wide array of industrial products from paints to lubricants. Most of this information is contained in insects and microbes that have not even been identified or catalogued. As these forests are destroyed for logging, cattle ranching and peasant farming, this genetic information is lost forever.

In the last 600 million years, there have been five great episodes of extinction on the planet. The most severe occurred at the end of the Permian period, 240 million years ago, when more than 80 percent of all marine-animal species vanished. The more famous episode of extinction occurred when the dinosaurs disappeared—probably because of the impact of a huge meteorite or comet—at the junction of the Cretaceous and Tertiary periods, 65 million years ago. In all five cases, animal species were affected more than plant species, and recovery to the preexisting level of biodiversity took from 10 million to 100 million years.

Today, both plant and animal species are vanishing at an extraordinary rate as a result of the rapid loss of tropical forests and their other habitats all over the planet. Harvard biologist E.O. Wilson has calculated conservatively that the global loss from tropical-forest degradation alone could be between 4,000 and 6,000 species a year, a rate 10,000 times greater than the natural rate of extinction prior to the appearance of human beings. If one adds the stress of fast climate change, an estimate of a 25 percent loss of planetary biodiversity in the next 100 years is quite realistic. This would rival four of the five previous mass extinctions on earth. From both a moral and practical point of view, it could be the single greatest calamity that human beings inflict on the planet.

✳ ✳ ✳

The above environmental problems can all be crudely described as large-scale and human caused, with long-term and often irreversible consequences, which is why they are often grouped together under the term "global change." Actually, however, they vary greatly in scale: climate change and ozone depletion involve genuinely global physical processes, while degradation of agricultural land and depletion of water supplies involve regional physical processes, although they may appear in places all over the planet.

These problems also vary in length of time over which they develop: for example, while a region can be deforested in only a few years, and serious environmental and social effects from this deforestation may be seen almost immediately, human-induced greenhouse warming will probably develop over many decades and may not have a truly serious impact

United Nations Photo 151.424/J.K. Isaac

**A fishing boat in the Bay of Bengal, where fish stock is disappearing.**

on people for a quarter of a century or more after the first detection of clear signs of warming. In addition, some of these problems (for instance, deforestation and degradation of water supplies) are much more advanced than others (such as greenhouse warming) and are already starting to cause serious social disruption. This variation in the amount of clear evidence for these various problems makes for great differences in our certainty about how severe they will become. For this reason, the uncertainties surrounding greenhouse warming are far greater than those for deforestation.

Finally, many of the above problems are interrelated. For instance, greenhouse warming may cause deforestation because it will move northward the ideal temperature and rainfall zones for many tree species (and unlike annual crops, trees do not adapt quickly enough to survive); it may make windstorms and wildfires worse; and it may allow pests and diseases to affect larger areas. The release of carbon from these dying forests would, of course, reinforce the greenhouse effect.

# Glossary of Key Terms

- **aquifer:** a water-bearing layer of permeable rock, sand or gravel capable of yielding considerable quantities of water to wells or springs

- **biodiversity:** the variety of the world's organisms, including their genetic makeup and their communities and interrelationships

- **biosphere:** the part of the world where life can exist; the term refers to living beings and their environment

- **chlorofluorocarbons (CFCs):** gases found in car air conditioners, coolants, aerosol products and solvents which, when released in the atmosphere, contribute to the breakdown of the ozone shield (see below)

- **desertification:** the process that turns land into arid waste or desert due to mismanagement or climate change

- **ecosystem:** a complex ecological community (e.g., a tropical forest) and its environment which together form a unit in nature

- **environmental degradation:** man-caused damage to the basic natural resources necessary for survival: water, soil, forests, the atmosphere

- **genetic engineering:** the alteration of genes, which control hereditary characteristics, by human intervention

- **global warming:** warming of the earth due primarily to the rapid buildup in the atmosphere of carbon dioxide and other "greenhouse gases" (see below) that trap the earth's heat

- **green revolution:** the achievement of higher food production in developing countries through the use of hybrid grain seeds and scientific methods of farming

- **greenhouse gas:** a gas in the atmosphere, which like the glass in a greenhouse, traps part of the earth's heat and produces a warming effect (carbon dioxide produces 50%; CFCs, 25%)

- **ozone depletion:** breakdown of the ozone shield (a thin layer of ozone gas molecules in the atmosphere) that can absorb damaging ultraviolet radiation and have major implications for global weather; CFCs greatly speed the process

# The Social Effects of Environmental Scarcity

There are three main sources of human-induced environmental scarcity. First, as discussed earlier, human activity can either reduce the quantity or degrade the quality of an environmental resource faster than the resource is renewed. Experts often refer to this process using terms similar to those used for money in a bank: the capital in a savings account and the interest or income it earns. Similarly, a renewable resource also has both "capital" and "income." The capital is the stock of the resource that generates a flow of income that can be tapped for human consumption and well-being. The topsoils on the Great Plains of America are capital that can be used to grow grain. A "sustainable" farming economy can be defined as one that leaves these soils—this capital—intact and undamaged so that generations of Americans can enjoy an undiminished stream of income in the form of food products. So, if the natural decay of crop residue adds about one hundredth of an inch of topsoil to farmed land a year (or about 3.25 tons/hectare), then sustainable farming should not produce an average soil loss greater than this amount.

The second main cause of environmental scarcity is population growth, which can reduce the amount of a renewable resource available per person. Over time, for example, a given flow of water or a given area of cropland might have to be divided among more and more people (as shown in Table 1).

Third, a change in the way a resource is distributed in a society can concentrate the resource in the hands of a few people, while the rest suffer extreme scarcity. Each society has, for example, rules and laws that define the limits and nature of ownership of things such as farmland. These "property rights" affect how resources are distributed among different groups in the society, and they often change as a result of large-scale development projects or new technologies that alter the relative values of resources. A particularly good example of this is the case of the Senegal River basin (discussed later), where a large dam dramatically raised the value of downstream farmland, which caused a powerful group to change the laws governing ownership of land to benefit themselves.

These three sources of environmental scarcity can operate singly or in combination and can have a variety of critical social effects, including declining food production, general economic stagnation or decline, displacement of population and the disruption of institutions and traditional social relations among people and groups. These social effects, in turn, are often interlinked, and sometimes reinforce each other. For example, economic uncertainty may lead to the flight of people with wealth and education, and such a money and brain drain in turn weakens universities, courts and financial institutions, all of which are absolutely necessary for a healthy economy.

## Agricultural Production

Experts often mention that constrained food output is the most worrisome possible result of environmental scarcity. As noted above, agricultural land is decreasing in availability and quality. For certain regions of the world, this may in time sharply limit food output, depending upon the ability of societies to increase the "intensity" of their agricultural practices, which means increasing output per hectare (commonly called the land's "yield") through strategies like mechanization, greater fertilization and irrigation, and new seeds.

But there are many other factors besides land availability that may influence agricultural production, including deforestation, depletion of water supplies, greenhouse warming, increased ultraviolet radiation from ozone depletion, and, eventually, decreased biodiversity. The serious forest degradation in the Philippines, described earlier, provides a good illustration of the impact on food output. Across the archipelago, logging and land clearing have accelerated erosion, changed regional hydrological cycles (the cycles of water between the atmosphere, land and plants that help determine local weather patterns) and decreased the land's ability to hold water during rainy periods. The resulting flash floods have damaged irrigation works while plugging reservoirs and irrigation channels with silt. This may seriously affect crop production. For instance, the Philippine government and the European Community wrote an integrated environmental plan for the still relatively unspoiled island of Palawan in the west of the Philippine archipelago. But the authors of the study found that only about half of the 36,000 hectares of irrigated farmland projected to be included in the "Plan for 2007" would actually be irrigable in that year because of the effects of decreased forest cover on water flows and irrigation capacity.

Greenhouse warming and climate change may also affect agricultural production, although this is a contentious issue. Coastal cropland in countries such as Bangladesh, Egypt and China is very vulnerable to ocean surges caused by big storms. Such events could become more common and devastating because global warming will cause sea levels to rise and could intensify hurricanes, cyclones and typhoons. The greenhouse effect will also change rainfall patterns and soil moisture; while this may benefit some agricultural regions, others will suffer, especially in poor countries. The World Resources Institute contends that "the impacts on agriculture could be double-edged: by altering production in the main food-producing areas, climate change could weaken our ability to manage food crises, and by making growing conditions worse in food-deficit nations, it could increase the risk of famine." Countries at special risk from climate change will be those—such as the nations of the Sahel in Africa—with an imbalance between population and food-growing ability and with little money to fund changes in their agricultural systems. As some areas become too dry to grow food, others, formerly dry, will suddenly have enough water; poor countries will not be able to afford the new dams, wells, irrigation systems, roads and storage silos that they need in order to adjust.

Many plants grow faster and larger in a warm environment rich in carbon dioxide, and they often use water more efficiently. But optimistic estimates of increased crop yields have been based on laboratory experiments under ideal growing conditions, including ideal amounts of soil nutrients and water. In addition, these estimates have ignored the influence on yields of more-frequent extreme climate events (especially droughts and heat waves), increased insect infestation and the decreased nutritional quality of crops grown in a carbon-dioxide-enriched atmosphere.

In general, the magnitude of climate change is less of a problem for poor countries than its rate. Around the world, human beings and their agricultural systems have adapted to differences in temperature far greater than the maximum warming predicted for the next 100 years. But the rapid rate expected for this change will produce new pressures on society at a time when it is already stressed by other population and resource problems. This change may be too fast and complex for societies that have limited buffering capacities.

Mexico, for example, is extremely vulnerable to climate change. Historically, environmental degradation—especially declining soil fertility—appears to have played a key role in the collapse of Mesoamerican civilizations, such as the Mayans. Today, large numbers of people are leaving the state of Oaxaca because of drought and soil erosion. In the future, global warming could produce a decrease of 40 percent in Mexican rain-fed agriculture, which, in combination with free trade (Mexico's trade advantage is in water-intensive fruits and vegetables) and the privatization of communal peasant lands, could bring great suffering and national conflict.

### Economic Decline

A very important potential social effect of environmental scarcity may be the further poverty it causes in already poor countries.

The ability of countries to produce wealth may be affected directly by environmental stress or indirectly by lower food output and population movements caused by environmental problems. Numerous examples are possible. The higher ultraviolet radiation caused by stratospheric ozone depletion is likely to raise the incidence of disease in humans and livestock, which may cost poor societies dearly. Logging for export markets—as in Southeast Asia, West Africa and Central America—results in a short-term economic gain for the country's elite, but the deforestation also greatly affects longer-term economic productivity. Increased runoff damages roads, bridges and other valuable infrastructure; extra siltation may destroy spawning grounds, and it can reduce the capacity of rivers to generate hydroelectric power, as well as their usefulness as transport routes for ships and barges. In addition, as forests are destroyed, wood becomes scarcer and more expensive, and it takes more of the household budget of poor families using it for fuel (in the Philippines, 70 percent of poor households use wood for fuel).

Farming is the source of a large share of the wealth generated in poor societies. It is not uncommon for nearly 50 percent of a country's gross national product (GNP, or the total output of a nation's goods and services in a specified period of time) to be generated by farming, while 60 percent or more of its population depends on agriculture for jobs. Food output has soared in many regions over the last decades because expanded irrigation, fertilizer use and new agricultural technologies have produced a "green revolution"* that has more than compensated for the declining soil fertility and depth. But some experts believe this economic relief will be short-lived. Jeffrey Leonard of the Overseas Development Council notes: "Millions of previously very poor families that have experienced less than one generation of increasing wealth due to rising agricultural productivity could see that trend reversed if environmental degradation is not checked."

In many poor countries the effects of land scarcity and degradation will probably become much clearer as the gains from green-revolution technologies are fully realized. But despite the extravagant claims of some commentators, a second green revolution of new agricultural technologies is not waiting in the wings to keep food productivity rising. Genetic engineering* may in time help scientists develop salinity-resistant, drought-resistant and self-fertilizing grains, but their wide use in poor countries is probably decades in the future.

Damage to the soil is already producing a harsh economic impact in some areas. Commenting on Indonesia, economist Robert Repetto of the World Resources Institute asserts:

> With erosion, farm output and income have fallen in some regions without major changes in farm practices; other farmers have been induced to change cropping patterns and input use; and in extreme cases, erosion has led to the complete withdrawal of land from cultivation. Farmers in the Citanduy Upper Watershed grow corn, upland rice and cassava on better soils. As erosion becomes more severe, rice is replaced by peanuts; and on nearly depleted soils only cassava is grown.

Unfortunately, measuring the actual cost of land degradation is not easy. Current national income accounts—statements of GNP—do not include measures of resource depletion. Repetto also notes: "A nation could exhaust its mineral reserves, cut down its forests, erode its soils, pollute its aquifers and hunt its wildlife to extinction—all without affecting measured income." Such weak measures of economic productivity reinforce the belief in governments of poor countries that environmental protection can only be achieved at the cost of economic growth; this belief, in turn, encourages societies to generate present income at the expense of their potential for future income.

After a careful analysis of soil types, cropping practices, logging and erosion rates in upland areas of Java in Indone-

**Chopping wood in Transkei, a daily act of resource depletion**

sia, Repetto concludes that the country's national income accounts "significantly overstate the growth of agricultural income in Indonesia's highlands." The true economic cost of soil degradation is "the present value of losses in farm income in current and future years." Taking lost future income into account, Repetto calculates the one-year cost of erosion in Indonesia to be $481 million, which is about 40 percent of the annual value of upland crop production. He concludes: "Nearly 40 cents in future income is sacrificed to obtain each dollar for current consumption." He also estimates that costs downstream from eroded hillsides, including the higher expense of clearing waterways and irrigation channels of silt, come to $30 million to $100 million a year.

In recent research on China, Vaclav Smil has estimated the combined effect of environmental problems on current economic productivity. The main economic burdens he identifies are reduced crop yields caused by water, soil and air pollution; more human sickness from air pollution; farmland loss because of construction and erosion; nutrient loss and flooding due to erosion and deforestation; and timber loss arising from poor harvesting practices. Smil calculates the current cost to be at least 15 percent of China's GNP. He is convinced that the toll will become much heavier during the coming decades.

## Population Displacement

Some observers have claimed that environmental degradation may produce vast numbers of "environmental refugees." For example, a sea-level rise due to global warming may drive people back from coastal and delta areas in Egypt; soil degradation and desertification may empty countries in the African Sahel as their populations move south; and Philippine fishermen may leave their ruined fishing grounds for the cities. But the term environmental refugee is misleading because it implies that environmental scarcity will be the direct and sole cause of refugee flows. Usually it will be only one of a large number of interacting physical and social fac-

Thomas Homer-Dixon

**Destitute Philippine peasants have migrated to Manila and built "houses" on a smoldering garbage heap—Smoky Mountain—which provides them with a meager livelihood selling bits of refuse.**

tors that may together force people from their homelands.

The term also does not distinguish between people who are fleeing due to genuine disaster or acute hardship, and those who are migrants for a variety of less urgent reasons. In general, migrants are motivated by a combination of "push" and "pull" factors, while refugees are motivated primarily by "push" factors. Additionally, many people do not have the financial resources to move easily; so those who leave as migrants are usually the skilled and semiskilled workers who can best afford it.

Research suggests that environmental refugees will appear only when there is a sudden and large environmental change. If environmental scarcity develops gradually, migration is more probable than refugee movement, and these migrants will tend to blend indistinguishably with migration streams due to other causes.

The northeast region of the Indian subcontinent provides a good example of population displacement arising from environmental scarcity. (See map, page 51.) Over the last three decades, as noted, land scarcity has been a key factor causing the large-scale movement of people from Bangladesh to the Indian state of Assam. In the future, people may be driven from Bangladesh by other environmental problems, including rising sea levels combined with cyclones (made worse by climate change) and terrible flooding due to deforestation in watersheds upstream on the Ganges and Brahmaputra rivers, which drain the subcontinent and empty through Bangladesh into the Bay of Bengal. Similarly, Vaclav Smil predicts that over the next decades Chinese in the tens of millions will try to move from the country's impoverished interior and northern regions, where water and fuelwood are desperately scarce and the land often badly damaged, to the booming cities along the coast. Bitter disputes will develop between these regions over water-sharing and migration.

### Disrupted Institutions and Social Relations

All societies are held together by a thick fabric of institutions, organizations, rules, customs and habitual behavior. This fabric includes branches of government like the court system, financial institutions like rural banking cooperatives, customary patterns of community government such as regular meetings of village chiefs, traditions of family organization, religious practices and institutions, and systems for land inheritance in the countryside. While changes in this fabric are often good for society, the social effects of environmental scarcity described above will frequently hurt people. Falling agricultural output will weaken rural villages through malnutrition and disease and by encouraging people to leave. Economic decline will corrode confidence in the national purpose and undermine financial, legal and political institutions.

And mass migrations of people into a region will drive down wages; shift relations between workers, peasants and landowners; and upset the long-standing balance of economic and political power among ethnic groups.

The effect environmental scarcity has on the state in developing countries deserves particular attention. The "state" is the set of institutions and people that manages the distribution of wealth and power within a society; it includes legislative bodies, the government's administrative bureaucracy (such as its departments of finance and foreign affairs), the court system, the police and the military. Research suggests that environmental scarcity can weaken the state by making it less able to administer its affairs. Environmental scarcity can also undermine the state's moral authority or "legitimacy" by making the state seem less fair and reasonable.

First of all, environmental scarcity sharply raises financial and political demands on government. For instance, it often requires huge spending on things like new dams and irrigation systems to compensate for water scarcity and fertilizer plants and reforestation programs to compensate for soil degradation. Environmental scarcity also drives up the number of "marginal" people who barely survive on the edges of society and who are desperate for government help. Around the world—in Peru, Brazil, the Sudan and India—the same process takes place. Unequal distribution of land and wealth combines with rapid population growth, failing farms, lack of water and lack of forests and fuel to cause poverty in the countryside. Many people are forced to move to cities where they demand food, shelter, transportation and jobs. In the Philippines, landless peasants come on boats from destitute islands like Negros and Leyte to the capital city of Manila. Right beside Manila's port they find their new home. They build their houses on top of an immense, squalid garbage dump, which locals call "Smoky Mountain" since it is perpetually smoldering from spontaneous combustion. The newcomers make their living sorting through the garbage and selling the scraps of plastic and metal they find.

In response to swelling urban populations, governments subsidize or freeze the prices of basic goods within cities like bread and rice, gasoline, electricity and bus transportation. Such price controls can discourage farmers and other entrepreneurs from producing these goods and cause companies and banks to make bad investments. Over time, this hampers economic growth, which hurts marginal groups, like those on Smoky Mountain, even more. In addition, as has been the case in countries as different as Kenya and Thailand, state interference in the economy shifts political and economic power within the society to a small urban elite, often consisting of just a few families, friends and allies of the country's rulers.

Simultaneous with all these events, the loss of renewable resources, from fish and fertile land to abundant forests, can reduce tax revenues to local and national governments. There is, then, the potential for a widening gap between demands on the state and its financial ability to meet these demands. Such a gap could in time boost frustration within a poor society, erode the state's legitimacy, and increase competition between cliques and factions within its elite as they struggle to protect their shares of the economic pie.

# Violent Conflict and
# Environmental Scarcity

If food production stagnates, if some developing societies slide further into poverty, if large numbers of people leave their homelands, and if institutions and social relations are disrupted, what kinds of conflict are likely to develop?

There is not much information on which to base an answer to this question. This may be partly because environmental and population pressures have not yet passed a critical threshold of severity in many poor countries. Also, environmental problems are very complex, and until recently there was little good research on environment-conflict linkages. But on the basis of the case studies in the project on "Environmental Change and Acute Conflict," three types of conflict seem most likely.

### Three Perspectives on Conflict

**Scarcity conflicts** are those one intuitively expects when countries calculate their self-interest in a world where the amount of resources is fixed, that is, in a world where the resource pie does not grow. Such conflicts will probably arise over three types of resources in particular: river water, fish and good cropland. These renewable resources are likely to spark conflict because their scarcity is increasing swiftly in some regions, they are often critical to human survival, and they can be physically seized or controlled.

The current controversy over the Euphrates River, discussed at the start of this book, illustrates how scarcity conflicts can arise. It is clear, though, that the problem of Euphrates water is tangled up with issues of territorial integrity and relations between government and ethnic minorities in both Syria and Turkey. Although water scarcity is a source of serious tensions between Syria and Turkey, and may produce interstate violence in the future, this dispute is not a pure example of a scarcity conflict. Truly pure examples may be impossible to find.

Experts in international relations who address the security implications of environmental scarcity usually emphasize the potential for scarcity conflicts. Yet research within the project on environmental change shows that these conflicts will not be the most common to arise from environmental stress. Indeed, ethnic disputes and corrosive conflict within countries deserve a greater portion of expert attention.

**Group-identity conflicts** are likely to arise from the large-scale movements of populations caused by environmental scarcity. As different ethnic and cultural groups are pushed together, people in these groups usually see themselves and their neighbors in terms of "we" and "they"; in other words, they will use the identity of their own group to judge the worth of other groups. Such attitudes often lead to bitter hostility and even violence. The situation in the Bangladesh-Assam region may be a good example of this process; Assam's ethnic strife over the last decade has apparently been catalyzed by immigration from Bangladesh. This case is discussed further below.

Growing population and environmental stresses in poor

UN Photo 154290/John Isaac

**People fleeing a drought-stricken area in Chad**

countries will undoubtedly lead to surging immigration to the industrialized world. Princeton political scientist Richard Ullman writes: "The image of islands of affluence amidst a sea of poverty is not inaccurate." People will seek to move from Latin America to the United States and Canada, from North Africa and the Middle East to Europe, and from South and Southeast Asia to Australia. This migration has already shifted the ethnic balance in many cities and regions of rich countries, and governments are struggling to contain a backlash against "foreigners." Such racial or ethnic strife will become much worse.

Although it seems probable that environmental scarcity will cause people to move in large numbers, which will in turn produce conflict, several qualifications are needed. First, refugees tend to have limited means to organize and to make demands on the government of the receiving society. Rather than overt violence, therefore, a common result will

# Figure 3. Some Sources and Consequences of Environmental Scarcity

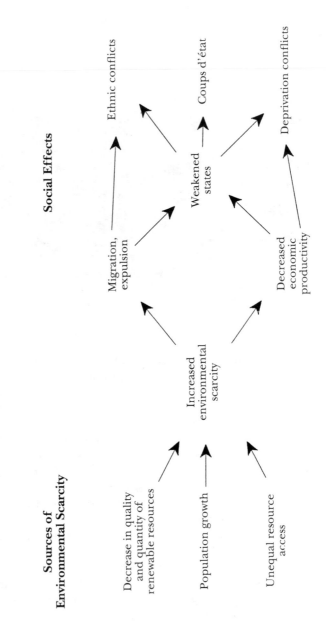

be silent misery and death, with little destabilizing effect. Second, government intervention plays a critical role in determining whether population displacement causes conflict; displaced groups often need the backing of a government (either that of the receiving society or of an external one) before they have sufficient power to cause conflict. And finally, it should be noted that migration does not always produce negative results. It can, for instance, ease labor shortages in the receiving society, as has been the case, for instance, in Malaysia. Countries as diverse as Canada, Thailand and Malawi demonstrate that societies often have a striking capacity to absorb migrants without conflict.

**Deprivation conflicts.** As poor societies produce less wealth because of environmental problems, their citizens will probably become increasingly angered by the widening gap between their actual standard of living and the standard they feel they deserve. The rate of change is key: the faster the economic deterioration, the greater the discontent. Lower-status groups will be more frustrated than others because elites will use their power to maintain, as best they can, the same standard of living despite a shrinking economic pie. At some point, the frustration and anger of certain groups may cross a critical threshold, and they will act violently against those groups perceived to be the agents of their economic misery or those thought to be benefiting from an unfair distribution of economic rewards in the society.

In general, experts on conflict within societies say that rebellion, revolution and insurgency are likely when (1) there are clearly defined and organized groups in a society; (2) some of these groups regard their level of economic achievement, and in turn the broader political and economic system, as wholly unfair; and (3) these same groups believe that all peaceful opportunities to achieve change are blocked, yet regard the balance of power within the society as unstable; in other words, they believe there are opportunities for overthrowing authority in the society.

Environmental scarcity helps produce both the second

and third of these conditions. A key social effect in poor countries is the disruption of institutions such as the state. Thus, environmental problems may not only increase the frustration and anger within poor societies (through increased deprivation); in addition, by weakening the state and other institutions, environmental problems may open up opportunities for angry groups to overthrow existing authority. These groups will also be more likely to challenge authority in a society if they have good leaders and are well-organized. Leaders help the members of their group believe that their situation should and can be changed. All of these factors are evident in the case study of the Philippines, in the next section.

## Case Studies

Three sources of environmental scarcity, identified earlier, are degradation and depletion of renewable resources, population growth and changes in resource distribution among groups. In this section, the case studies will show how these three sources operate singly or in combination to produce the types of conflict just discussed. The case studies are drawn from the work of the research project on environmental change. They suggest that group-identity and deprivation conflicts will be particularly common outcomes of environmental scarcity (see Figure 3 on page 48).

### • Bangladesh-Assam

Population growth by itself will be a key source of social stress and conflict in some cases. Bangladesh, for instance, does not suffer a critical loss of agricultural land from erosion or nutrient depletion because the normal floods of the Ganges and Brahmaputra rivers lay down a layer of silt every year that helps maintain the fertility of the country's vast plains. But the UN predicts that Bangladesh's current population of 115 million will grow to 235 million by the year 2025. Cropland is already desperately scarce at about 0.08 hectares per capita, but since all of the country's good agri-

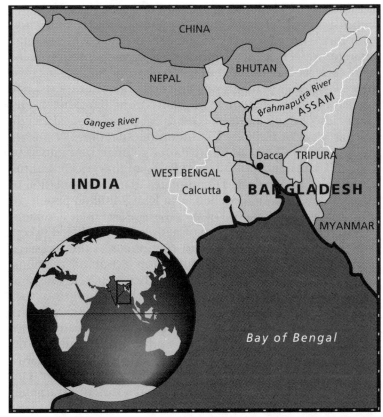

cultural land has already been exploited, population growth will cut this figure in half by 2025. Land scarcity and the brutal poverty and social turmoil it produces have been made worse by repeated severe floods (perhaps aggravated by deforestation in the watersheds of the region's major rivers) and the susceptibility of the country to cyclones.

As noted previously, over the last four decades these stresses have caused millions of people to migrate from Bangladesh, formerly East Pakistan, to neighboring areas of India. Detailed data are scarce; the Bangladeshi government is reluctant to admit that there has been a significant exodus because the problem has become a major source of friction with India. But one of the researchers on the environmental-change project, Sanjoy Hazarika, an investigative journalist who reports for *The New York Times* from South Asia, has pieced together demographic information and experts' estimates. He concludes that migrants from Bangladesh and their descendants have increased the population of neighboring areas of India by 15 million people, of which only 1 million or 2 million can be attributed to migration caused by the 1971 war between India and Pakistan that resulted in the creation of Bangladesh. He estimates that the population of the state of Assam has swelled by at least 7 million people, of Tripura by 1 million, and of West Bengal by about 7 million.

This enormous movement of people has produced sweeping changes in the politics and economies of the receiving regions. It has altered land distribution, economic relations, and the balance of political power between religious and ethnic groups, and it has triggered serious intergroup conflict, for example between the Lalung tribe and Bengali Muslim migrants in Assam. In the neighboring Indian state of Tripura, the original Buddhist and Christian inhabitants now make up only 29 percent of the state's population, with the rest consisting of Hindu migrants from Bangladesh. This shift in the ethnic balance precipitated a violent insurgency between 1980 and 1988 that was called off only after the government agreed to return land to dispossessed Tripuris and stop the influx. But because the migration has continued, this agreement is in jeopardy.

India offers certain attractions for potential migrants from Bangladesh. The standard of living is often much better; furthermore, Indian politicians have often encouraged Bangladeshis to enter India in order to expand certain segments of the voting population. The region's conflicts must also be seen in the light of history. During the colonial period, for instance, the British used Hindus from Calcutta to administer Assam. The Assamese are therefore particularly sensitive to their growing loss of political and cultural control in the region as more people migrate from Bangladesh. While this historical context is important, it does not obscure the powerful influence of environmental scarcity behind these conflicts.

● **The Senegal River Basin**

Elsewhere in the world, population growth and damage to renewable resources often encourage large-scale development projects that can shift resource access. This can produce dire conditions for poorer and less powerful groups whose claims to resources are opposed by elites, and this may lead to violence. A good example is the dispute that began in West Africa in 1989 between Mauritania and Senegal over the Senegal River valley which defines their common border.

Senegal has fairly abundant agricultural land, but much of it suffers from high to severe wind and water erosion, loss of nutrients, salinization and soil compaction caused by agriculture. The country has an overall population density of 38 people per square kilometer and an annual population growth rate of 2.7 percent, giving a doubling time of about 25 years.

Except for the valley of the Senegal River along its southern border and a few oases, Mauritania is largely arid desert and semiarid grassland. Although its population density is very low, at approximately 2 people per square kilometer, the population growth rate is 2.8 percent. This combination of factors led the FAO in a 1982 study to include both Mauritania and Senegal in their list of "critical" countries whose croplands cannot support their current and projected populations without a large increase in agricultural inputs, like fertilizer and irrigation.

The floodplains fringing the Senegal River are broad and fertile, and have long supported a productive economy—farming, herding and fishing—based on the river's annual floods. During the 1970s, a serious drought and the prospect of a chronic food shortfall in the region encouraged the region's governments to seek international financing for the high Manantali Dam on the Bafing River tributary in Mali and the Diama salt-intrusion barrage near the mouth of the Senegal River between Senegal and Mauritania. These dams were designed to regulate the river's flow to produce hydropower, expand irrigated agriculture and provide river transport from the Atlantic Ocean to landlocked Mali. But there were other social and economic consequences that the international experts and regional governments did not foresee.

Senegal's population is mainly black; but prior to 1989 there was a small but vigorous class of white Moor shopkeepers throughout much of the country. Mauritania, in contrast, is dominated by a white Moor elite, and there has been a long history of racism by these Moors toward their non-Arab black compatriots. This racism is deeply resented by black Africans across the border in Senegal.

As anthropologist Michael Horowitz has found, anticipation of the new dams sharply increased land values along the river where irrigation could be installed. The Moor elite in Mauritania suddenly became interested in title to this land and rewrote the legislation governing its ownership. This effectively stripped black Africans of their rights to continue farming and herding along the Mauritanian bank of the Senegal River, as they had done for generations.

In the spring of 1989, tensions in the river basin triggered an explosion of attacks by blacks against Moors in Senegal, which led to attacks by Moors against blacks in Mauritania. Within a few weeks, almost all of the 17,000 shops owned by Moors in Senegal had been pillaged or destroyed, several hundred people had been killed in ethnic violence in both countries, and thousands had been injured. Nearly 200,000 refugees fled in both directions across the border, and the two countries were nearly at war. The Mauritanian regime used this occasion to activate new land legislation, declaring the black population along the Mauritanian portion of the river basin to be Senegalese, stripping them of their citizenship and seizing their properties and livestock. Many of these blacks were forcibly expelled to Senegal, and some launched cross-border raids to retrieve expropriated cattle.

This case study illustrates the interaction of the three sources of human-induced environmental scarcity identified earlier. Agricultural shortfalls, caused in part by population pressures and degradation of the land resource, encouraged a large development scheme. These factors together raised land values in one of the few areas in either country offering the potential for a rapid move to modern, high-input agriculture. The result was a change in property rights and resource distribution, a sudden increase in resource scarcity for an ethnic minority, expulsion of the minority and ethnic violence.

### ● The Jordan River Basin

The water shortage on the occupied West Bank of the Jordan River offers a similar example of how population growth and excessive resource consumption and degradation can promote unequal resource access. While figures vary, one of the project's researchers, Miriam Lowi of Princeton University, estimates that the total amount of renewable fresh water annually available to Israel is about 1,950 million cubic meters (mcm), of which 60 percent comes from groundwater (i.e., aquifers), and the rest from river flow, floodwater and waste-water recycling. Current Israeli demand, including the needs of settlements in the occupied territories and Golan Heights, is around 2,200 mcm. The annual deficit of some 200 mcm is covered by overpumping aquifers. As a result, the water table in some parts of Israel and the West Bank is dropping significantly (Israeli experts estimate from 0.2 to 0.4 meters per year), thus causing the salinization of wells and the infiltration of sea water from the Mediterranean. Moreover, Israel's population is expected to increase from the present 4.6 million to 6.5 million in the year 2020, even without major immigration from the countries of the former Soviet Union. Based on this population growth projection, the country's water demand could exceed 2,600 mcm by 2020.

Two of the three main aquifers on which Israel depends lie principally underneath the West Bank, with their waters draining into Israel proper. Consequently, nearly 40 percent of the groundwater Israel uses originates in occupied territory. In order to protect this important source for Israeli consumption, the Israeli government has strictly limited water use on the West Bank. Of the 650 mcm of all forms of water annually available in that territory, Arabs are allowed to use only 125 mcm. Israel restricts the number of wells Arabs can drill in the territory, the amount of water Arabs are allowed to pump, and the times at which Arabs can draw irrigation water. The differential in water access on the West Bank is marked: on a per capita basis, Jewish settlers consume about four times as much as Arabs.

Arabs are not permitted to drill new wells for agricultural purposes, although the Mekorot (the Israeli water company) has drilled more than 30 wells for settlers' irrigation. Arab agriculture in the region—especially in areas such as Jericho—has suffered because many Arab wells have become dry or saline as a result of deeper Israeli wells drilled in proximity. This, combined with the confiscation of agricultural land for settlers and other Israeli restrictions on Palestinian agriculture, has encouraged many West Bank Arabs to abandon farming. Those who have done so have become either unemployed or day laborers within Israel.

The Middle East as a whole faces increasingly grave and tangled problems of water scarcity, and many experts believe these will affect the region's security. Water was a factor contributing to tensions preceding the 1967 Arab-Israeli war, and the war gave Israel control over most of the Jordan basin's water resources. While "water wars" in the basin are possible in the future, they seem unlikely given the military predominance of Israel. More probably, in the context of age-old ethnic and political disputes, water shortages will aggravate social tensions and unrest within societies in the

# Figure 4. Some Sources and Consequences of Environmental Scarcity in the Philippines

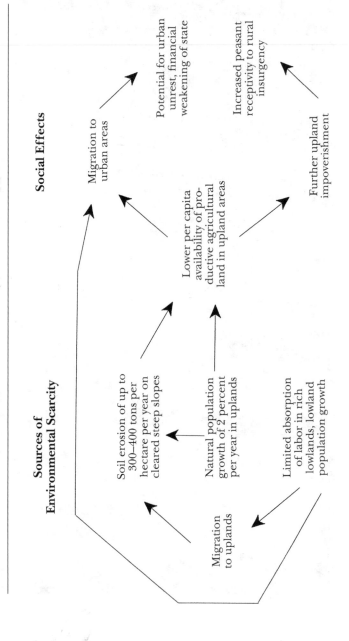

**Sources of Environmental Scarcity**

**Social Effects**

Migration to uplands

Soil erosion of up to 300–400 tons per hectare per year on cleared steep slopes

Natural population growth of 2 percent per year in uplands

Limited absorption of labor in rich lowlands, lowland population growth

Lower per capita availability of productive agricultural land in upland areas

Migration to urban areas

Potential for urban unrest, financial weakening of state

Further upland impoverishment

Increased peasant receptivity to rural insurgency

Jordan River basin. In recent congressional testimony, political scientist Thomas Naff noted that "rather than warfare among riparians [i.e., countries that border or straddle rivers] in the immediate future, what is more likely to ensue from water-related crises in this decade is internal civil disorder, changes in regimes, political radicalization and instability."

## • The Philippines

In many parts of the world, a somewhat different process is under way: unequal resource access combines with population growth to produce environmental damage. This can lead to economic deprivation that spurs insurgency and rebellion.

In the Philippines, Spanish and American colonial policies left behind a grossly unfair distribution of land. Since the 1960s, the government, often in collusion with the country's powerful landowners, has promoted the modernization and further expansion of large-scale agriculture in the country's low-lying coastal plains. This was done to raise the output of grain for domestic consumption and of cash crops such as coconuts, pineapples, bananas and sugar to help pay the country's massive foreign debt. In general, this has held static or lowered the labor demand per hectare (the "labor-intensity") of agricultural production. With a population growth rate of 2.3 percent and with little rural industrialization to absorb excess labor, agricultural wages have fallen, and the number of poor agricultural laborers and landless peasants has swelled. Economically desperate, millions of people have migrated to shantytowns in already overstressed cities, such as the Smoky Mountain squatters camp in Manila, while millions of others have moved to the least productive and often most ecologically vulnerable territories, especially steep hillsides away from the coastal plains (see Figure 4).

In these uplands, settlers use fire to clear forested land or land that has been previously logged by commercial companies closely associated with landowning, government and military elites. The settlers bring with them little money or knowledge to protect their fragile ecosystems, and their small-scale logging, production of charcoal for the cities and slash-and-burn farming often cause horrendous environmental damage, including water erosion, landslides and changes in the hydrological cycle. This has set in motion a cycle of falling food production, clearing of new plots and further land degradation. Even marginally fertile land is becoming hard to find in many places, and economic conditions are often hazardous for the peasants.

The country has suffered from serious internal strife for many decades. But two of the project's researchers, Celso Roque, the former undersecretary for the environment of the Philippines, and his colleague Maria Garcia, have concluded that resource stress appears to be an increasingly powerful force driving the Communist-led insurgency. Some senior Philippine politicians have reached the same conclusion. Daniel Lacson, the governor of the province of Negros Occidental under President Aquino, identifies two sources of poverty and injustice behind the insurgency: the accumulation of land in the hands of a few who have failed to deal with the problems of the poor; and land degradation that affects small farmers and is not alleviated by government action.

The upland insurgency is motivated by the economic deprivation of the landless agricultural laborers and poor farmers displaced into the hills where they try to eke out a living from failing land; it exploits opportunities for rebellion in the country's peripheral regions, which are largely beyond the effective control of the central government; and it is helped by the creative leadership of the cadres of the New People's Army and the National Democratic Front. During the 1970s and 1980s, these groups found upland peasants very receptive to revolutionary ideology, especially where the repression of landlords and local governments left them little choice but to rebel or starve. The revolutionaries have built on local understandings and social structures to help the peasants define their situation and focus their discontent.

60

## Other Similar Cases

Processes similar to those in the Philippines are strikingly common around the planet. Population growth and unequal access to land force huge numbers of people either into cities or onto marginal lands. In the latter case they cause major environmental damage and become chronically poor. Eventually, they may be the source of persistent unrest and insurgency, or they may migrate from these marginal regions yet again, stimulating ethnic conflicts elsewhere or urban unrest. The short but violent "Soccer War" between El Salvador and Honduras in 1969 involved just such a combination of factors.

As William Durham of Stanford University has shown, changes in agricultural practice and land distribution in El Salvador since the mid-nineteenth century concentrated poor farmers in the country's uplands. Over the years these farmers developed some understanding of land conservation, but inevitably their increasing numbers on very steep hillsides caused serious deforestation and erosion. A natural population growth rate of 3.5 percent further aggravated land scarcity, and as a result many people migrated to neighboring Honduras where there was more open land. Tension gradually increased between the newcomers and the local Hondurans, and the eventual expulsion of the migrants from Honduras started the war. Durham notes that the competition for land in El Salvador that led to this conflict was not addressed after the war; it was a powerful factor behind the country's subsequent civil war; and it was a central issue in El Salvador's arduous peace negotiations.

Similarly, in South Africa, the white regime's past *apartheid* policies concentrated millions of blacks in the "homelands," created by the South African government in some of the country's least productive and most ecologically sensitive territories, where population densities were made worse by high natural birthrates. In 1980, rural areas of the Ciskei homeland had a population density of 82 persons per square kilometer, while the surrounding predominantly white Cape

**61**

Province of South Africa had a rural density of 2. Homeland residents have little money and few resource-management skills and are the victims of undemocratic, corrupt and abusive local governments. Sustainable development in such a situation is impossible, and wide areas have been completely stripped of trees for fuelwood, grazed down to bare dirt and eroded of topsoil. A 1980 report concluded that nearly 50 percent of Ciskei's land was moderately or severely eroded, and nearly 40 percent of its pasturage was overgrazed.

This loss of the resource base, combined with a lack of alternative means of employment and social breakdown due to the stresses induced by apartheid, has produced an economic crisis in the homelands. This, in turn, has caused high levels of migration to South African cities, which are still unable to integrate and employ these migrants adequately, due to the social and economic structures remaining from apartheid. The result is the rapid growth of squatter settlements and illegal townships that are rife with violence that threatens the country's move to full democracy.

# Long-term Security Implications

Can poor countries respond effectively to environmental scarcity to prevent the processes described above? If some of them cannot, what will this mean for national and international security over the long term?

## *Supply and Demand of Human Ingenuity*

Some optimists argue that it is not so much environmental scarcity per se that is important as whether or not people are harmed by it. Human suffering might be avoided if political and economic systems provided the incentives and ingenuity that allow people to reduce or eliminate the harmful effects of scarcity. To address this argument, more information is needed about the factors that affect the supply and demand of human ingenuity in response to environmental scarcity.

The combined or "aggregate" data on world food production might seem to justify optimism. Between 1965 and 1986, many poor regions suffered serious environmental problems, including erosion, salinization and land loss to urbanization. Yet global production of cereal grains increased 3 percent a year, meat and milk output grew by 2 percent annually, while the growth rate for oil crops, vegetables and legumes was 2.5 percent. At the regional level, increased food production kept ahead of population growth, except in Africa, and shortfalls were eased by exports from rich countries with huge surpluses. One might therefore conclude that poor countries have sufficient resources and ingenuity, with occasional assistance from grain-exporting countries, to respond to environmental problems.

But such aggregate figures hide large differences in food availability among and within poor countries. Moreover, these statistics are not as promising as they once were: many developing countries have already reaped most of the green revolution's potential benefit, and the rate of increase in global cereal production has declined by over 40 percent since the 1960s. For three successive years—from 1987 through 1989—global cereal consumption exceeded production. While bumper grain crops were again harvested in 1990, carryover stocks can be rapidly exhausted, and as few as two years of bad harvests in key agricultural regions could produce a global food crisis.

Over the long term, the ability of poor countries to respond well to the effects of environmental scarcity on agriculture will depend on the complex interactions within each society of the factors indicated in Figure 2. Particularly important factors are the society's dominant land-use practices, land distribution and market mechanisms within the agricultural sector. The latter is especially important today, since numerous poor countries are giving up state control over their markets, reducing government spending and allowing more foreign investment.

According to most economists, in a market economy that gives accurate prices to all traded goods and services, environmental scarcity will encourage conservation, technological innovation and substitution of more-abundant resources for scarce ones. Princeton University economist Julian Simon, in particular, has an unwavering faith in the capacity of human ingenuity to overcome scarcity when spurred by self-interest. Many economists point to the success of the green revolution, which they claim was driven in part by market forces, since, among other things, it involved the substitution of relatively cheap petroleum resources (in the form of fertilizer) for increasingly degraded, and therefore more expensive, agricultural land. This powerful argument supports the policies for liberalizing markets in poor countries advocated by international financial and lending institutions, such as the International Monetary Fund and the World Bank. But these policies may not be an effective response to environmental scarcity in the future.

### Cornucopians and Neo-Malthusians

Experts in environmental studies now commonly use the labels "cornucopian" for optimists like Simon and "neo-Malthusian" for pessimists like the scientists and authors Paul and Anne Ehrlich. Cornucopians do not worry much about protecting the stock of any single resource because of their faith that market-driven human ingenuity can always be tapped to allow the substitution of more-abundant resources to provide the same service. Simon writes: "There is no physical or economic reason why human resourcefulness and enterprise cannot forever continue to respond to impending shortages and existing problems with new expedients that, after an adjustment period, leave us better off than before the problem arose."

The neo-Malthusian label comes from the eighteenth-century economist Thomas Malthus, who claimed that severe hardship was unavoidable because human population always grows much faster than food production. Neo-Malthusians are much more cautious than cornucopians in their assessment of human ability to push back resource limits. For renewable resources, they often make the distinction between "capital" and "income" discussed above; they contend that

**65**

human economic activity should leave environmental capital intact and undamaged so that future generations can enjoy a constant stream of income from these resources.

Historically, however, cornucopians have been right to criticize the idea that resource scarcity places fixed limits on human activity. Time and time again, human beings have circumvented scarcities, and neo-Malthusians have often been justly accused of "crying wolf." But in assuming the experience of the past pertains to the future, cornucopians overlook seven factors.

## Cornucopian Oversights

First, whereas serious scarcities of critical resources in the past usually appeared singly, now multiple scarcities interact with each other to produce synergistic effects. An agricultural region may, for example, be simultaneously stressed by degraded water and soil, greenhouse-induced changes in rainfall and increased ultraviolet radiation. The total impact of these interacting problems may be much greater than the sum of their separate impacts. Moreover, thresholds and extreme events make the future highly uncertain for governments and corporations faced with economic and investment decisions. Finally, as numerous resources become scarce at the same time, it is harder to identify substitution possibilities that serve the same purposes at costs that prevailed when scarcity was less severe.

Second, in the past the scarcity of a given resource usually increased slowly, allowing time for social, economic and technological adjustment. But human populations are much larger and the activities of people consume, on a global average, many more resources than before. This means that critical scarcities often develop much more quickly: whole countries are deforested in a few decades; most of a region's topsoil can disappear in a generation; and critical ozone depletion may occur in as little as 20 years.

Third, today's consumption has far greater momentum than in the past because of the size of the population of consumers and the sheer quantity of material consumed by this population. The countless individuals and groups making up societies are heavily committed to certain patterns of resource use; and the ability of markets to adapt may be sharply limited by these special interests.

These first three factors may soon combine to give a new and particularly severe character to global environmental scarcity: the populace will face multiple resource shortages that are interacting and unpredictable, that quickly grow to crisis proportions, and that will be hard to address because of powerful commitments to certain patterns of consumption.

The fourth reason that cornucopian arguments may not apply in the future is that the free-market price mechanism is a bad gauge of scarcity, especially for resources held commonly by all people, such as a favorable climate and productive seas. In the past, many such resources seemed endlessly abundant; now they are being degraded and depleted, and it is becoming evident that their greater scarcity often has tremendous bearing on a society's well-being. Yet this scarcity is at best reflected only indirectly in market prices. In addition, people often cannot buy or sell in a market in which they have an interest, because they either lack the money or are distant from the market in space or time. For instance, poor coastal communities in the southern Indian state of Kerala see their local fish catches dwindling year-by-year because of overfishing offshore; yet they do not have the money or ability to buy fish in international markets. Similarly, future generations who will suffer from today's overfishing cannot buy fish in today's markets either. In such cases the true scarcity of the resource is not reflected by its price.

The fifth reason is an extension of a point made earlier: market-driven adaptation to resource scarcity is most likely to succeed in wealthy societies. Abundant reserves of money, knowledge and talent help economic actors invent new technologies, identify possibilities for conservation, and make the transition from old to new patterns of production and consumption. Yet many of the countries facing the most serious

**67**

environmental problems in the coming decades will be poor; even if they have efficient markets, lack of money and know-how will keep them from responding well to these problems.

Sixth, cornucopians have an outdated faith in human beings' ability to unravel and manage the complex processes of nature. There is no fundamental reason to expect that scientific and technical ingenuity can always overcome all types of scarcity. People may not have the mental capacity to understand adequately the complexities of environmental and social systems. Even with this capacity, it may simply be impossible, given the physical, biological and social laws governing these systems, to reduce all scarcity or repair all environmental damage. Moreover, scientific and technical knowledge must be built incrementally—layer upon layer—and its diffusion into the broader society often takes decades. Technological solutions to environmental scarcity, if indeed any are to be found, may arrive too late to prevent catastrophe.

Seventh, future environmental problems, rather than inspiring the wave of ingenuity predicted by cornucopians, may actually reduce the supply of ingenuity available to a society. Two kinds of ingenuity are key to dealing with environmental scarcity. "Technical ingenuity" is needed for the development of, for example, new agricultural and forestry technologies that compensate for environmental degradation. "Social ingenuity" is needed for the creation of institutions and organizations that buffer people from the effects of degradation and provide the right incentives for people to develop new technologies. The development and distribution of new grains adapted for dry climates and eroded soils, of alternative cooking technologies to compensate for the loss of firewood, and of water conservation technologies depend on an intricate and stable system of markets, laws, financial agencies and educational and research institutions. Not only are poor countries undersupplied with these social resources, their ability to create and maintain them will be weakened by the very environmental stresses they need to address.

Cornucopians often overlook the role of social ingenuity in producing and maintaining the complex legal and economic climate in which technical ingenuity can flourish. Governments, bureaucrats and local community leaders must be clever "social engineers" to design and implement good adaptation strategies and market mechanisms. Unfortunately, though, multiple, interacting, unpredictable and rapidly changing environmental problems will dramatically increase the complexity and urgency of the setting within which government and community leaders must operate; and they will also generate increased "social friction" as elites and interest groups struggle to protect their shares of economic and political power. The ability of governments and others to respond to the increased demand for social ingenuity by being good social engineers will likely go down, not up, as these stresses increase.

When it comes to the poorest countries on this planet, therefore, one should not invest too much faith in the potential of human ingenuity to respond to environmental problems once they have become severe. The most important of the seven factors above is the last: growing population, consumption and environmental stresses will increase decision-making complexity and social friction. These will in turn reduce the ability of leaders in poor countries to intervene as good social engineers in order to chart a sustainable development path and prevent further social disruption. Neo-Malthusians may underestimate human adaptability, but as time passes their analysis will become ever more telling.

### The Future

What does this mean for the future of poor countries faced with severe environmental scarcity? Such a country will follow one of four paths. Most optimistically, the country may have enough technical and social ingenuity to permit substitution of relatively plentiful resources for scarce resources or to foster social adaptation to scarcity. For example, on the technical side, the country might use or im-

**69**

port genetic-engineering technology to develop crops that can grow on agricultural land that is salinized from overirrigation; on the social side, peasants and landowners might learn to share scarce resources more fairly.

If this is not possible, the country might be able to "uncouple" itself from dependency on its own environmental resources by making goods and services that can be traded on the international market for the environmental resources it needs (such as grain). The uncoupling might, in fact, be achieved by rapidly exploiting the country's environmental resources, which allows it to accumulate enough money, industrial equipment and educational resources to permit a shift to other forms of production. For instance, the country might use the income from overlogging its forests to fund a modern university system that trains electrical engineers and computer specialists for a high-technology industrial sector.

If the country is unable to do either of these things, the government's instruments to manage society—its financial, bureaucratic and political machinery—may eventually be so damaged by environmentally induced social stress that chronic protest and violence arise. The country may then fragment as peripheral regions come under the control of renegade authorities and "warlords." Such a weakening of government control over outer regions is under way in the Philippines, and environmental scarcity may be a strong contributor to this process.

Finally, and most pessimistically, the state may avoid fragmentation by becoming a "hard" regime, that is, by violating human rights and resorting to authoritarian measures against internal opposition. As political philosopher William Ophuls notes, "scarcity in general erodes the material basis for the relatively benign individualistic and democratic politics characteristic of the modern industrial era; ecological scarcity in particular seems to engender overwhelming pressures toward political systems that are frankly authoritarian by current standards...." A hard regime may also launch military attacks against neighboring countries to distract its people's attention from their rising grievances. Research suggests that the correlation between internal strife and attacks against neighboring countries depends on the nature of the government and the kind of internal strife it faces. For example, highly centralized dictatorships are most likely to start wars with other countries when threatened by revolutionary actions and strikes. In comparison, less-centralized dictatorships are most prone to war when threatened by guerrilla attacks and assassinations. External aggression may also result after a new government comes to power through civil violence: governments born of revolution, for example, are particularly good at mobilizing their citizens and resources for war.

## Which Way to Go?

Whether or not a given country will follow the third or fourth path above depends in part on the level of remaining resources and wealth in the country's economic system that can potentially be seized by the state for its own uses. This suggests that the countries with the highest probability of becoming hard regimes (and potential threats to their neighbors) in the face of environmental and economic stress are large, relatively wealthy developing countries that are dependent on a declining environmental base, such as Nigeria, Indonesia, Brazil, India and China. In contrast, countries such as Bangladesh, Haiti, the Philippines, Ethiopia and the Sudan are more likely to fragment.

If many developing countries evolve in the direction of extremism, the security interests of rich countries may often be directly threatened. Of special concern here is the growing gap between rich and poor nations that may be widened by environmental scarcity. Economist Robert Heilbroner notes that new revolutionary regimes "are not likely to view the vast difference between first class and cattle class with the forgiving eyes of their predecessors." And these nations may be heavily armed, as the continued proliferation of nuclear and chemical weapons and ballistic missiles suggests. Such regimes, Heilbroner claims, could be tempted to use nuclear blackmail as a "means of inducing the developed world to transfer its wealth on an unprecedented scale to the under-developed world." Political scientist Richard Ullman, how-ever, believes this concern is overstated. Third World nations are unlikely to confront rich countries violently because of the "superior destructive capabilities of the rich." In light of the discussion in this article, one might conclude that envi-ronmental stress and the social disruption it produces will so weaken the economies of poor countries that they will be unable to build sizeable armed forces. But the developed world would surely be unwise to rely on poverty and disorder in the developing world for its security.

## Conclusion: What Can Be Done?

The global environmental situation is critical, and it will soon cause social disruptions that may threaten national and international security. But these processes can be altered. A great deal of knowledge exists about what people are doing to the environment, about which kinds of social systems damage the environment least, and about how best to adapt to ecological stress.

Over the last few years there have been some heartening moves in the right direction by national governments and international agencies. For example, countries and chemical companies have moved aggressively to limit and eventually stop the production of the CFCs that destroy the ozone layer. The international agreements to do this include novel ways of transferring funds from rich to poor countries to help them move to technologies that do not use CFCs. This may be a precedent for transfer mechanisms to help poor countries develop energy technologies that produce less car-bon dioxide.

International financial agencies, like the World Bank, the Asian Development Bank and the International Monetary Fund, have become much more aware of the intricate rela-tionship between economic development and environmental scarcity. They are more cautious about funding large-scale, centralized development projects, like hydroelectric dams, and they are subjecting all their financing to more careful environmental review. In addition, they will soon provide substantial funding for sustainable development projects around the world.

Without question, the most important recent environmental event was the United Nations Conference on Environment and Development (Unced) in Rio de Janeiro, Brazil, in June 1992. This was the largest and most inclusive international negotiation in history. In arduous meetings leading up to the conference, delegates from rich and poor nations tried to work out preliminary agreements to protect biodiversity, control carbon-dioxide emissions, and slow deforestation. Some environmentalists were disappointed by the final results of the conference: the United States refused to sign the biodiversity treaty; the carbon-dioxide agreement was significantly weakened (again as a result of U.S. pressure); and some poor nations objected to the deforestation agreement. Nonetheless, the Rio conference was a milestone in humankind's efforts to meet the environmental challenge. It brought worldwide attention to the crisis, and it showed politicians and policymakers that good environmental policy can be a source of political authority and moral legitimacy.

### Recommendations

The research project on "Environmental Change and Acute Conflict" has produced some recommendations for action in rich and poor countries. Some of these are very general and will need elaboration by experts and negotiation among societies. Others are quite specific. Importantly, some of the recommendations do not require huge financial investments but may nonetheless make a great difference to the future well-being of our world.

In response to resource scarcity in the past, leaders have usually sought to increase the supply of scarce resources rather than decrease demand. In the future, action on the demand side of the scarcity problem will be particularly crucial.

While circumstances differ across the developing world, in general poor countries should act to limit population growth, they must try to implement a fairer distribution of land and wealth within their societies, and they should encourage sustainable economic growth to provide employment for surplus labor.

In the short term, the leaders of many poor countries see a trade-off between economic growth and environmental protection, and this encourages overuse and degradation of environmental resources. But the real trade-off is between short-term unsustainable prosperity and long-term growth potential. This is an answer to the common argument from poor countries that rich countries are using environmental issues to deny them the opportunity to grow: it is in the developing world's self-interest to prevent environmental decline. Policymakers in both rich and poor countries must therefore change their understanding of development to emphasize the true value of natural resources and the importance of efficiencies in production as a way of conserving resources.

In general, environmental problems, from water scarcity to declining fish stocks, require regional or "ecosystemwide" solutions rather than policies based at the national level. This point also applies to the management of the social effects of environmental scarcity: for instance, if large-scale migration is caused by environmental change, countries should develop regional accords to limit and manage it.

Some successful solutions of regional environmental problems may be transferable from one region to another. A good example is the International Joint Commission between Canada and the United States that manages, among other things, disputes over water. But the many international treaties and institutions that govern water supplies around the world are also inflexible. For instance, the existing water agreements between the United States and Mexico, and between Egypt and Sudan, allocate fixed amounts of water, whereas allocation based on a proportion of available water would be a more reasonable method when there is rapid climate change.

Population stabilization requires collaboration between rich and poor countries. Wealthy countries can usefully supply financial aid and technology to slow population growth, although the United States...sharply curtailed its support for such programs [under the Reagan and Bush Administrations]. Unfortunately, some poor countries often resist family planning for political and ideological reasons. Islamic fundamentalists and the Roman Catholic Church, for example, are opposed to birth control. In addition, leaders of some poor countries perceive that a large population will enhance military security.

Rich countries can also support land reform in developing countries, reduce the debt burden of these countries, and promote, through aid and transfers of technology, rural industrialization projects. Poor nations under pressure from banks and international lending agencies to pay their foreign debts often use their best lands to grow cash crops for export. As people are displaced from these lands, governments and development agencies can work to provide them with jobs. Ideally, land reform combined with labor-intensive rural industries will boost incomes and stem the flow of people into ecologically vulnerable areas and into cities that are increasingly unmanageable.

Rich countries can help poor countries overcome their lack of expertise on environmental management. In the developing world, inequality in national expertise translates into inequality in national power. Countries with advantages in technical expertise can dominate their neighbors; they have the edge in negotiations over resource management; and this can lead to misperceptions, rivalry and conflict. Charles Okidi, an international lawyer at Moi University in Kenya, notes that there are only three Ph.D.-level hydrologists in his country. The addition of four per country in East Africa would greatly improve prospects for regional water management. It is cost-effective for rich countries to provide funds for the training of environmental experts in the developing world, including hydrologists, soil and agricultural scientists, foresters, demographers, energy-systems engineers and fisheries specialists. If research and teaching centers are adequately equipped and staffed, the brain drain from poor to rich countries could be stemmed. Moreover, networks of such centers established across national boundaries could start wider cooperation among a region's countries.

Wealthy countries can also help groups working for environmental causes in poor countries by supplying the basic communication and data-processing technologies they need. In the Philippines, 2,500 nongovernmental organizations work on environmental research, education and political action at all levels of society. Their effectiveness can be multiplied by aid to buy personal computers for word processing, hard disks for data storage and printers and modems for distributing their information. Rich countries should promote the work of these groups within and across national boundaries.

Beyond these policy recommendations, certain changes of attitude are needed. This was one of the main lessons of Unced. Rich countries must recognize that global environmental problems are, in important ways, problems of global fairness. Deep poverty in the developing world often causes activities and strengthens economic structures that harm the environment. Moreover, as poor countries industrialize, they will need massive financial and technological help to leap over the environmentally damaging technologies and practices that were used by today's rich countries to industrialize.

Poor countries also argued at the Rio meeting that environmental degradation is often driven, not by large populations, but by the high consumption and waste production of wealthy countries. Poor countries are not the main causes of, yet may be most vulnerable to, the large-scale environmental problems getting most attention in the developed world, like greenhouse warming and ozone depletion. For all these reasons, widespread environmental scarcity may force humankind at last to deal with the wrenching economic differences that plague the planet. The UN will play a vital role in this effort.

The evidence presented here shows there are key links between environmental scarcity and global security. If rich and poor countries want to reduce conflict and social turmoil in tomorrow's world, they should spend more on preventing environmental scarcity and less on maintaining their often huge military establishments. This would be a crucial step toward a sustainable, fair and peaceful global society for all.

# Talking It Over

*A Note for Students and Discussion Groups*

This issue of the HEADLINE SERIES, like its predecessors, is published for every serious reader, specialized or not, who takes an interest in the subject. Many of our readers will be in classrooms, seminars or community discussion groups. Particularly with them in mind, we present below some discussion questions—suggested as a starting point only—and references for further reading.

## Discussion Questions

Are you convinced that environmental scarcity can lead to severe conflict? Why, or why not?

Are scarcities of renewable resources more likely to produce severe social stress and conflict than scarcities of nonrenewable resources? If so, why?

What is the role of environmental scarcity as a cause of conflict? Is it best seen as, for example, a trigger that releases accumulated nonenvironmental social pressures (political, economic, ethnic, etc.), or as a powerful underlying stress on society?

How would people in developing countries see the issues discussed in this HEADLINE SERIES? What are the primary interests of the leaders of poor countries? Do these interests promote or impede solutions to environmental problems?

Do you think the public's current concern for environmental issues is simply a fad that will be pushed aside by other concerns in the future?

Are the social stresses and conflicts considered in this book really a matter for serious attention in rich countries? Are the security interests of the developed countries really threatened? How might people in rich countries be convinced that they should commit resources to dealing with these problems?

## READING LIST

Brown, Janet Welsh, ed., *In the U.S. Interest: Resources, Growth, and Security in the Developing World.* Boulder, Colo., Westview, 1990. An edited collection of articles on the security threats posed by environmental degradation to societies considered strategically vital to the U.S. national interest.

Clark, William, et al., "Managing Planet Earth." *Scientific American,* September 1989. A special issue that offers a comprehensive survey of the technical challenges of managing human interaction with the biosphere.

Durham, William, *Scarcity and Survival in Central America: The Ecological Origins of the Soccer War.* Stanford, Calif., Stanford University Press, 1979. A pioneering treatment of how population growth, land degradation and skewed land distribution can interact to precipitate severe conflict.

Ehrlich, Paul, and Ehrlich, Anne, *The Population Explosion.* London, Hutchinson, 1990. A fast-paced argument about the dangers of global population growth by two of the leading proponents of the neo-Malthusian view.

Gurr, Ted, "On the Political Consequences of Scarcity and Economic Decline." *International Studies Quarterly,* March 1985. A discussion of the long-term results of environmental scarcity on social stability by one of the preeminent researchers of the causes of revolution and civil violence.

Heilbroner, Robert, *An Inquiry into the Human Prospect.* New York, Norton, 1980. A classic survey by a leading economist of the daunting resource and economic crises facing humankind in the next decades.

Homer-Dixon, Thomas, "On the Threshold: Environmental Changes as

Causes of Acute Conflict." *International Security,* Fall 1991. A research agenda and theoretical framework for environment-conflict research.

Leonard, Jeffrey, ed., *Environment and the Poor: Development Strategies for a Common Agenda.* New Brunswick, N.J., Transaction, 1989. A collection of essays that describe strategies for sustainable development in different ecosystems around the world.

Mathews, Jessica Tuchman, "Redefining Security." *Foreign Affairs,* Spring 1989. A recent and widely read argument for expanding the concept of "security" to include nonmilitary threats to society.

Ophuls, William, *Ecology and the Politics of Scarcity: A Prologue to a Political Theory of the Steady State.* San Francisco, Calif., Freeman, 1977. A treatment of problems of political order in situations of chronic resource scarcity.

Simon, Julian, *The Ultimate Resource.* Princeton, N.J., Princeton University Press, 1981. A vigorous statement of the "cornucopian" position on human response to resource scarcity.

Westing, Arthur, ed., *Global Resources and International Conflict: Environmental Factors in Strategic Policy and Action.* New York, Oxford University Press, 1986. A volume of edited papers on various impending shortages of renewable resources—including water, food and fish—and their implications for international security.

World Commission on Environment and Development, *Our Common Future.* New York, Oxford University Press, 1987. The famous report of the Brundtland Commission that popularized the concept of "sustainable development."

World Resources Institute, et al., *World Resources 1992–93.* New York, Oxford University Press, 1992. An invaluable reference book on many aspects of environmental change, resource consumption and population growth; includes many tables with the latest statistical data.